Field G  *to*

Butterflies

of

Seychelles

Their Natural History and Conservation

By

James M. Lawrence

Paintings by

Malinda Crafford-Venter

 **SSP**

SIRI SCIENTIFIC PRESS

Authors, if you would like to publish with us please email your proposed titles
(at any stage of preparation) to:
books@siriscientificpress.co.uk
or contact us via our website
(we specialize in high quality, rapid production and short print run titles
that would tend to be overlooked by larger publishers)

ISBN 978-0-9574530-9-8
Published by Siri Scientific Press, Manchester, UK
This and related titles are available directly from the publisher at:
http://www.siriscientificpress.co.uk

Printed and bound in the UK

Contents

Right: Forest on Mahé

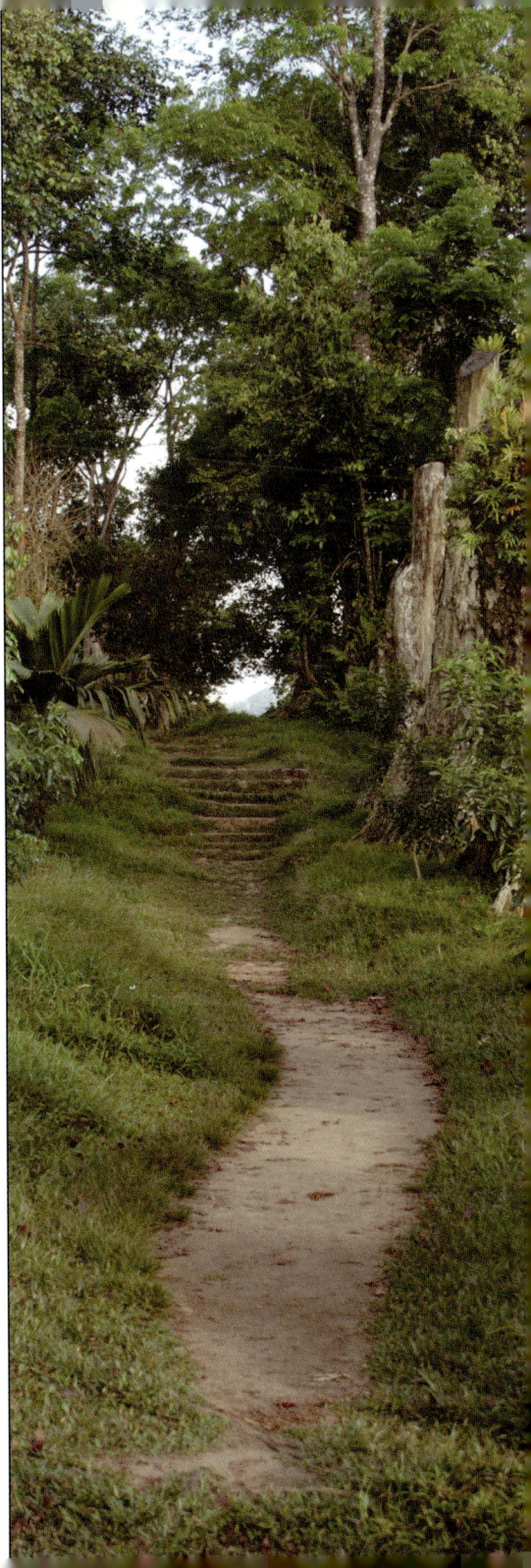

Foreword

The very word Seychelles conjures up a vision of paradise, of beautiful islands bathed in turquoise seas. These islands also have a mystique, the central group being an isolated fragment of ancient Gondwanaland. These inner granitic islands are graced by many outer islands formed on a base of coral reefs, and hence known as coralline islands. Much of the beauty of Seychelles is due to a fascinating set of animals, plants and fungi, partly of African origin, partly Asian, and then those that have evolved on the islands – the endemics.

Among the fauna are the butterflies, well-known as a group for their beauty and gracefulness. Perhaps less-often realised is that they are important sentinels of landscape change. Like all tropical islands worldwide, the Seychelles is under pressure from humans which are transforming the landscapes at an unprecedented rate. The butterflies on the islands are also under pressure and their population levels are indicative of deteriorating and, in some cases, improving conditions.

To be able to observe butterflies in their own right or to use them in conservation assessment, a first step is to be able to identify them correctly. Here, James Lawrence provides a stunning field guide with beautiful and classic original paintings by Malinda Crafford-Venter of all the Seychelles' butterflies. This enables naturalists and scientists alike to be able to identify species and subspecies with ease. They also present information on the biology and conservation status of all these butterflies.

This truly excellent book is a landmark publication on an important component of the Seychelles fauna, and is a must for all enthusiasts and professionals interested in the fauna and conservation of these intriguing islands.

Michael Samways PhD, FRSSAf, MASSAf

Professor and Chair,
Department of Conservation Ecology and Entomology,
Stellenbosch University,
South Africa

Acknowledgements

This book would not have been possible without valuable help from the following people:

D. Penney of Siri Scientific Press for editorial guidance and for publishing this book.

M.F. Keeley and J. Henwood of Cousine island, Seychelles for logistical support in Seychelles during field work; R. Russell and S. Russell for logistical support in the United Kingdom while doing museum work.

T. Larsen, M. Williams, D. Bernaud, O. Yata and J. Gerlach for providing information on species' taxonomy and biology.

J. Chainey of the Natural History Museum, London, United Kingdom; J. Hogan of the Oxford University Museum of Natural History, United Kingdom; R.C. Stebbings of the University Museum of Zoology, Cambridge, United Kingdom; M. Kruger of the Ditsong Museum of Natural History, South Africa.

S. Woodhall, P. Mazzei, D. Currie (davecurrie666@hotmail.com), A. Cosson, S. Tremeges (photos.twinlions.fr) and Premaphotos for providing photographs of living butterflies.

M. Samways for writing the foreword to this book. L. Zimmerman, D. Lawrence and P. Lawrence for help with book design, layout and proof reading.

Below: Desroches

D. Lawrence

Introduction

Insects form the most species-rich group of land animals. Following increased awareness of the vital functions they perform in many ecosystems (Samways et al. 2010a), their conservation has assumed considerable importance. Butterflies are the best known and most popular of the insect groups. They are aesthetically appealing and easy to identify by the non-specialist, making them of central interest for ecotourism and environmental education. Furthermore, due to their intimate relationship with plants, they are frequently used as indicators of ecosystem and habitat quality in monitoring programmes (Nelson 2007).

Tropical islands worldwide are losing indigenous biodiversity at an increasing rate from various, often synergistic, impacts (Primack 2006). The impacts caused by habitat loss (Samways 2005), in combination with other threats, in particular invasive alien species (Davies 2009), may lead to a 'meltdown' in the ecological character of an island (O'Dowd et al. 2003). This is particularly relevant for Seychelles invertebrate biodiversity, with its high levels of endemism (Gerlach 2008), and with the islands having undergone widespread historical anthropogenic habitat degradation (Stoddart 1984).

Assessing the conservation status of species has become a critical aspect of monitoring trends in biodiversity conservation at both national and global levels (Zamin et al. 2009). A fundamental aspect to achieving this is accurate identification of a species. To date, no book summarising all known Seychelles butterfly information, especially one illustrating every species, has been available. As most information is in scientific works and monographs, and therefore not easily accessible to the general public or to biologists, a fully illustrated guide was considered necessary.

The Seychelles butterfly fauna consists of both widespread cosmopolitan and endemic species. Several of the endemics are rare, while some are thought to be extinct. Many of these butterflies are known only from few or single sightings or captures. From the outset it was decided to include all butterfly species listed as occurring in Seychelles, including those that have only been rarely recorded.

The information presented in this book not only allows the reader to identify all known Seychelles butterfly species, but also provides background to the biology and conservation of these invertebrates. As will become evident, our knowledge of Seychelles butterflies is poor and fragmentary at best. This book will hopefully act as a starting point in advancing our knowledge of the natural history and conservation of these insects. The butterfly fauna of Seychelles is dynamic, with changes in species composition occurring due to the presence of ephemeral populations and species' extinctions, as well as the infrequent appearances of vagrant individuals. By placing on record the information presented in this book, any future changes in the butterfly fauna may then be seen in a historical perspective.

This book consists of three sections. The first section deals with a description of the Seychelles Islands. Here, the geography, climate and vegetation of the islands are covered. The second section discusses butterfly biology, starting by introducing butterfly taxonomy to the reader. Egg, larval and adult morphology are described, followed by a brief look at intraspecific variation and mimicry in adult butterflies. Seychelles butterfly biogeography is then covered, followed by a discussion on the conservation of Seychelles butterflies. The final section presents the systematic account, and deals with the identification and biology of the 36 species so far recorded on these islands.

Below: Cousine
Right: Wetland, Mahé

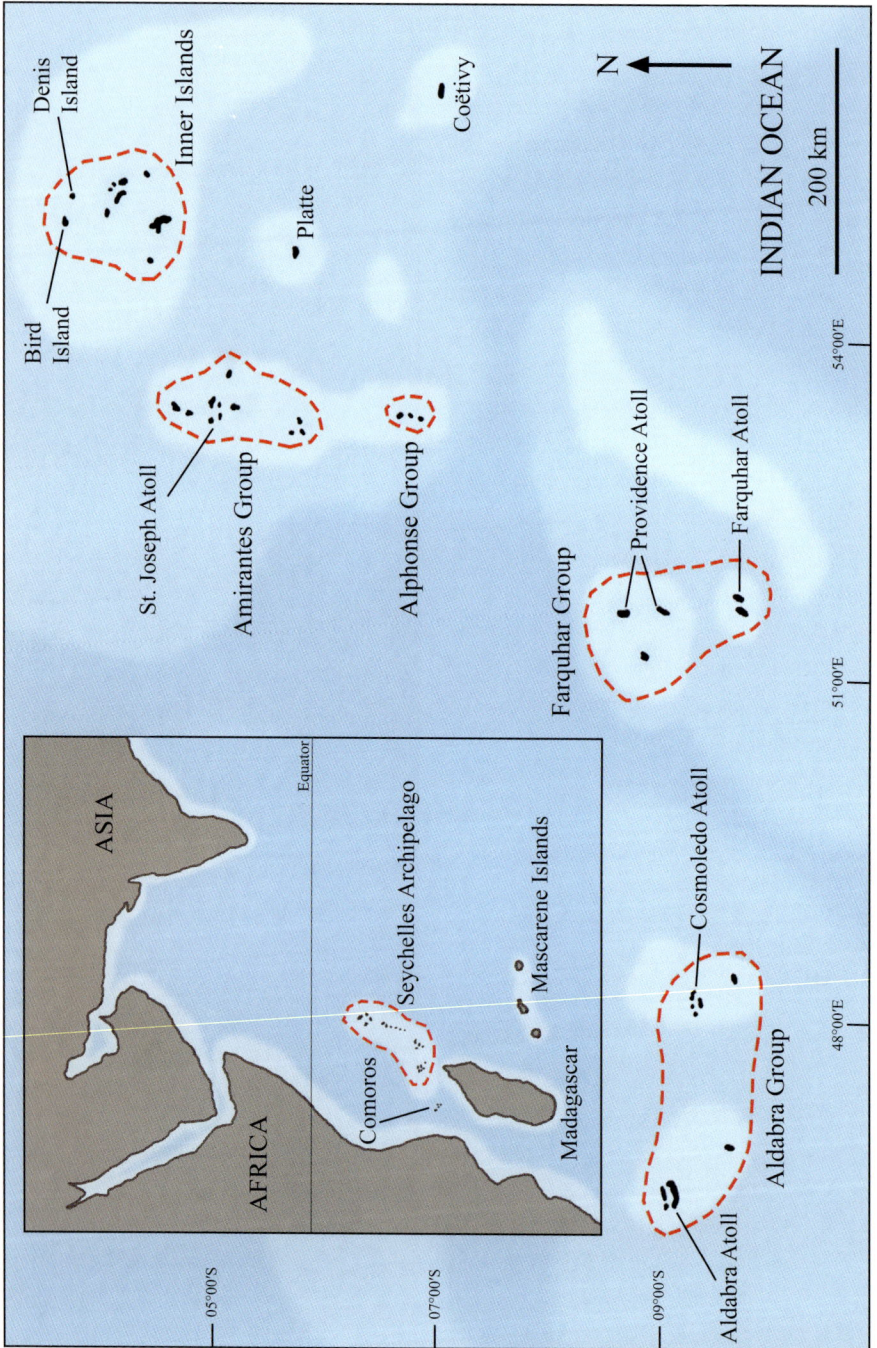

N

INDIAN OCEAN

200 km

Denis Island

Bird Island

Inner Islands

Platte

Coëtivy

St. Joseph Atoll

Amirantes Group

Alphonse Group

Farquhar Group

Providence Atoll

Farquhar Atoll

ASIA

AFRICA

Equator

Seychelles Archipelago

Comoros

Mascarene Islands

Madagascar

Cosmoledo Atoll

Aldabra Atoll

Aldabra Group

05°00'S

07°00'S

09°00'S

48°00'E

51°00'E

54°00'E

5

Seychelles Islands

Geography

The Seychelles Archipelago falls within the Malagasy subregion, comprises 115 islands and can be divided into two main island types: the granitic island group and the coral islands, which consist of the Amirantes, Alphonse, Farquhar and Aldabra groups.

The granitic islands are made up of 40 islands scattered across 400,000 km² of sea in the western Indian Ocean. These islands were part of the super-continent Pangaea, and are the only mid-oceanic islands that were not formed from coral or volcanic action. Approximately 200 million years ago, continental drift tore Pangaea apart, splitting the super-continent into Laurasia (North America, Europe and Asia) to the north and Gondwanaland (South America, Australasia, Antarctica, Africa and the Indian subcontinent) to the south. The Seychelles granitic rocks are approximately 650 to 750 million-years-old.

Approximately 125 million years ago, Madagascar, Seychelles and India broke away from Gondwanaland. Madagascar became an island around 90 million years ago, with Seychelles splitting from the Indian

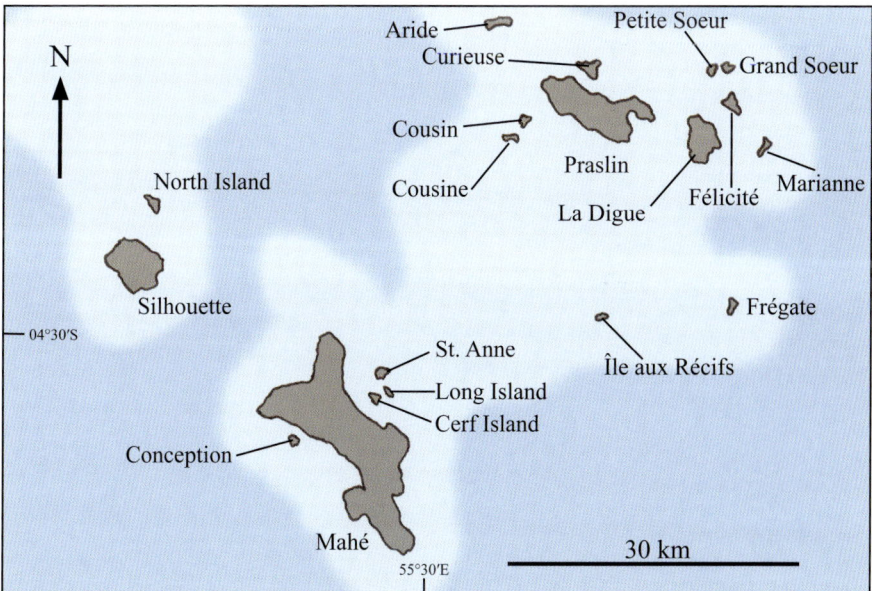

Above: Granitic Group
Opposite: Seychelles Archipelago; **Inset map**: Malagasy subregion

6

subcontinent approximately 65 million years ago (Braithwaite 1984).

Until as recently as 10,000 years ago the granitic Seychelles were a huge single landmass estimated to be about 300,000 km^2 (Camoin et al. 2004). Today, all that remains are the peaks of the highest mountains. The combined forces of sea-level change, erosion and coral growth weight have submerged all but 250 km^2 of this landmass. The highest point of these islands is on Mahé at 905 m a.s.l..

Though superficially similar to the granitic islands, North Island and Silhouette are much younger and consist of volcanic syenite rock around 63 million years in age. For convenience, these two islands are normally grouped together with the main granitic islands.

The 75 coral islands and atolls make up 47% of the land area of Seychelles, and are further divided into the low-lying and high coral islands.

Just to the north of the granitic islands lie the coral islands of Bird and Denis. Theses two islands, together with the granitic islands, are known as the Inner Islands.

The Amirantes, Alphonse and Farquhar island groups form a linear chain of islands to the south-west of the granitic islands. Along with the islands of Platte, Coëtivy, Denis and Bird, these three island groups are considered to be low-lying coral islands rising no more than 2 to 3 m a.s.l.. These low-lying islands are the youngest Seychelles islands, thought to have emerged about 4000 years ago when sea-levels dropped as a result of a change in ocean currents.

The Aldabra island group consists of Aldabra Atoll, Cosmoledo Atoll, Assumption

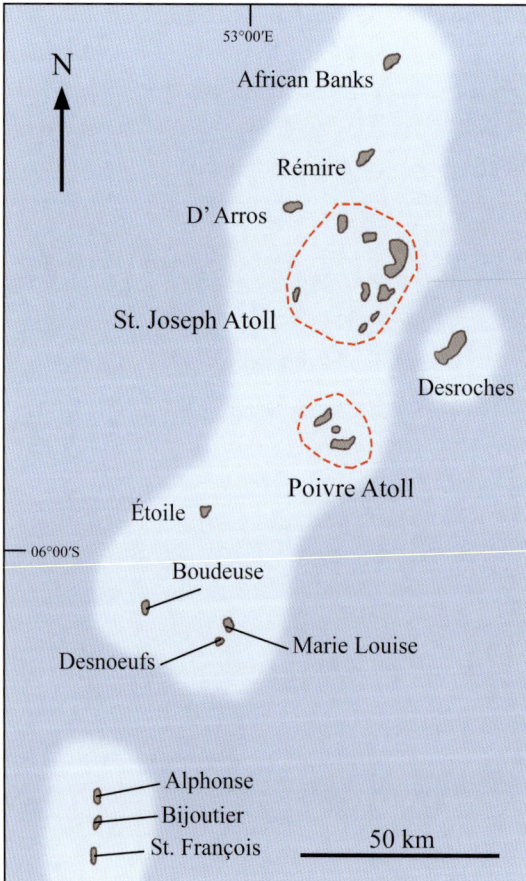

Left: Amirantes and Alphonse Groups

Above: Aldabra Group
Right: Farquhar Group

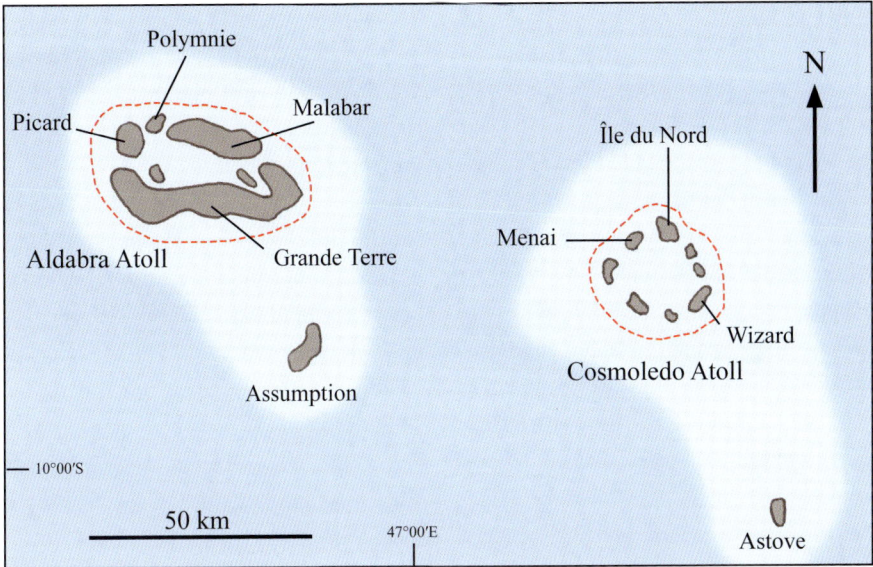

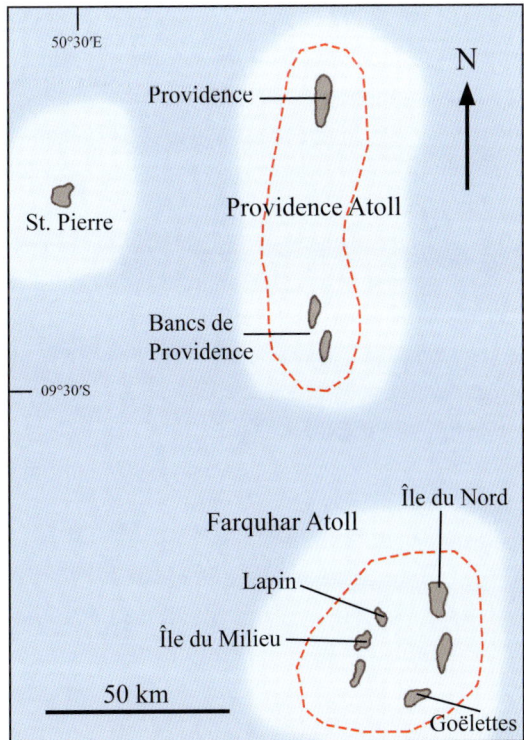

and Astove. These islands are the high coral islands, as they rise up to 8 m a.s.l.. Aldabra is made up of four main islands, Picard, Polymnie, Malabar and Grand Terre, and a number of smaller islets, enclosing a vast lagoon.

Together, the islands of Aldabra Atoll have a land area of more than 15,000 ha. The atoll is approximately 35 km long and 14 km wide. Over the long course of geological time Aldabra Atoll has undergone many submersions and emergences due to changes in sea-level, with the last emergence occurring approximately 125,000 years ago.

Climate

The following brief outline of the Seychelles climate is based on Walsh (1984). Although the Seychelles has a humid tropical climate in that annual rainfall exceeds 700 to 800 mm, and mean monthly temperatures are generally always above 20 °C, some important differences exist between the islands. These spatial variations in climate are due to both the altitudinal contrasts of the high granitic and low coral islands, and the vast expanse of ocean covered by these islands.

There are two main seasons in Seychelles. In the Southern Hemisphere winter (May to October), the SE Trade Winds extend over the whole of the western Indian Ocean. In the Southern Hemisphere summer (December to March), the NW Monsoon extends over the Seychelles islands. During the transition months of April and November, winds tend to be light and variable.

The temperature records for Port Victoria (Mahé) and Aldabra Atoll may be regarded as representative for sea-level locations across Seychelles. Mean annual temperatures are 26.6 °C at Port Victoria, and 27.0 °C on Aldabra. In common with other equatorial regions, annual variations are small, being approximately 2 °C on Mahé and 3 °C on Aldabra. Diurnal temperature ranges in Seychelles (3.6 °C at Port Victoria and 5.7 °C at Aldabra) are much lower than in continental equatorial areas, because oceanic influences reduce maximum temperatures and increase minimum temperatures. On the mountainous granitic islands, temperature decreases as altitude increases, with upland temperatures being about 3 to 4 °C cooler than those at sea-level.

In the granitic islands, altitude and aspect strongly affect mean annual rainfall. Rainfall increases with altitude and also tends to be higher on north-facing slopes. At sea-level on Mahé, the average total annual rainfall is 2500 mm, increasing to 5000 mm at higher altitudes

On the coral islands, annual rainfall varies significantly, decreasing markedly south-westwards through the islands. Mean annual rainfall on the north-eastern islands of Bird and Denis is twice that of the islands of Aldabra and Assumption. The Aldabra group is the driest, with an average total annual rainfall of approximately 1200 mm. Annual rainfall on the Amirantes and Alphonse island groups is intermediate, reflecting their central position along the island chain.

Rainfall also varies seasonally, and is closely linked to the SE Trade Winds and NW Monsoon periods. The SE Trade Winds bring dry weather to the Seychelles, whereas the NW Monsoon period and transition months experience high rainfall.

Seychelles is north of the normal path of tropical cyclones in the western Indian Ocean. They are rare in the granitic Seychelles. However, cyclones have been recorded in the south-western islands of Aldabra and Assumption, although they are infrequent.

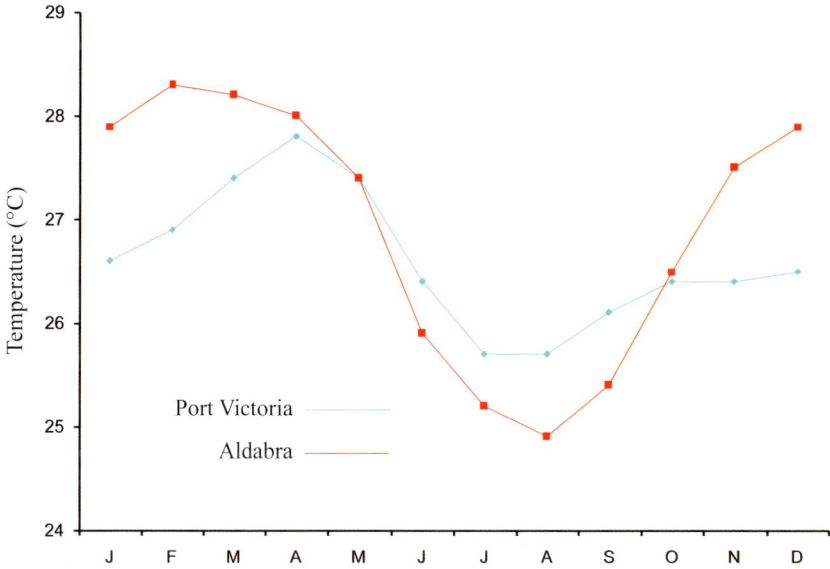

Above: Average monthly temperature (°C) (data from Walsh 1984)

Below: Average monthly rainfall (mm) (data from Walsh 1984)

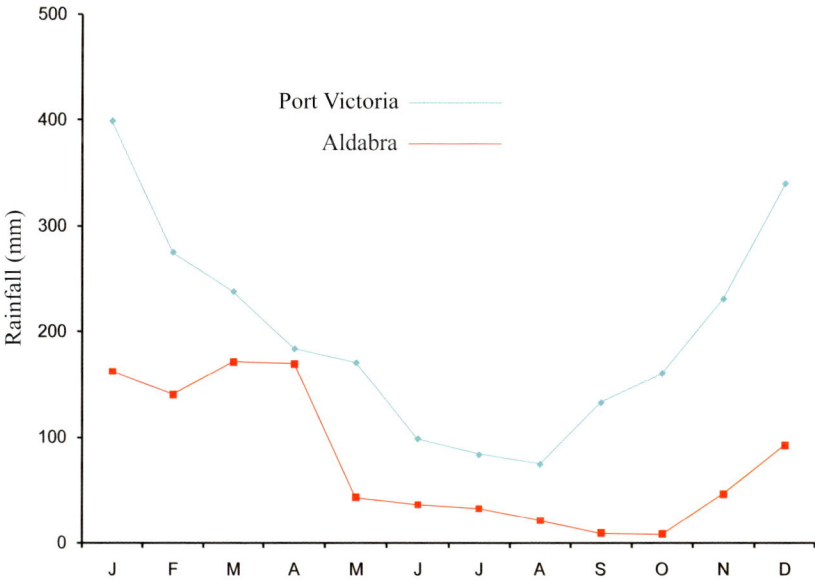

Vegetation

The following brief description of the islands' vegetation is summarised from Piggott (1968), Procter (1984a,b), Robertson (1989) and Hill & Currie (2007). The vegetation of Seychelles has changed considerably since they were first colonised. Originally the granitic islands were clothed with a dense covering of forest vegetation, while originally the coral island vegetation was dominated by widespread plant species that were dispersed on ocean currents and were tolerant to salt spray and drought.

Three main vegetation types are recognised on the larger granitic islands: 1) Moist forest; 2) Dry forest; 3) Coastal vegetation. These vegetation types are best divided by altitude.

Moist forest: This forest type extends between 500 to 900 m a.s.l., where the annual rainfall is higher than at the coast. These high moist forests are often dominated by many indigenous and endemic trees such as *Northea hornei*, *Dillenia ferruginea*, *Timonius sechellensis*, *Eugenia wrightii* and *Erythroxylum sechellarum*.

Dry forest: This forest type extends between 100 to 500 m a.s.l., and is characterised by less-abundant moisture than the moist forest type. Many of these forests have been replaced by plantations of *Sandoricum koetjape*, *Camellia sinensis* and *Khaya senegalensis*. On the larger islands the less-disturbed forests are dominated by palms, most notably the famous Coco de Mer, *Lodoicea maldivica*. Another well-known and interesting tree found in this biotope type is *Medusagyne oppositifolia*, commonly known as the Jellyfish tree. This endemic tree was thought to be extinct in the 1960s before being rediscovered on Mahé in the 1970s. This species has been placed in its own family, the Medusagynaceae.

Coastal vegetation: This vegetation type is found below 100 m a.s.l. on the coastal plateaux. Most of the original coastal plateau vegetation has been cleared and cultivated. The plateaux were historically dominated by trees such as *Calophyllum inophyllum* and *Terminalia catappa*. Formerly, extensive wetland and mangrove swamps were found on these low-lying areas.

The vegetation of the smaller granitic islands has more in common with that of the coral islands than that of the large granitic islands. Many of these islands were historically used for coconut cultivation and are now overgrown with alien vegetation. However, much vegetation-based restoration is taking place on many of these islands, with the aim of restoring the vegetation to what it was formerly thought to be. Such projects have been prominent on the islands of Cousine, Cousin, Aride and Frégate (Shah 2001, 2006; Henri et al. 2004; Samways et al. 2010b).

Much of the original vegetation on the low-lying coral islands has been lost to guano mining and coconut plantations. Coconuts, with re-growth of *Pisonia grandis* and *Morinda citrifolia* dominate the vegetation of these islands. Beach-crest vegetation consists mostly of *Scaevola sericea*, *Cordia subcordata*, *Guettarda speciosa* and *Tournefortia argentea*. Mangroves are rare on these low-lying coral islands.

The high coral islands of the Aldabra group are much older than the low-lying coral islands and therefore have a higher number of endemic floral species. Except for Aldabra Atoll, much of the vegetation on these islands has been destroyed by guano mining. Mangroves are extensive on these islands. *Pemphis acidula* scrub is widespread and abundant on Aldabra forming pure stands up to 6 m tall. Other abundant plant species include *Canthium bibracteatum*, *Sideroxylon inerme*, *Euphorbia* species and *Ficus* species. *Sporobolus virginicus* and *Sclerodactylon macrostachyum* form extensive grasslands on the coastal areas.

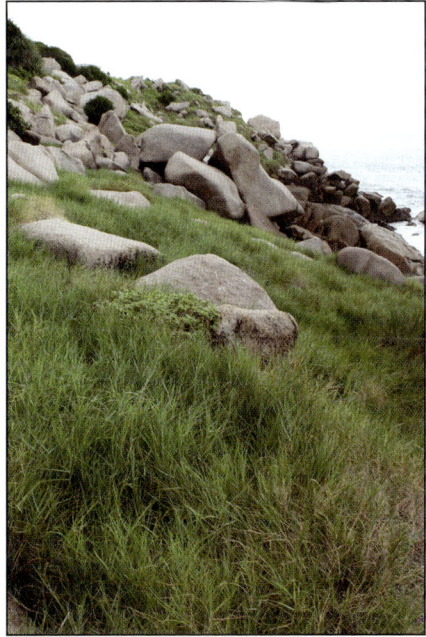

Above: Grassland vegetation
Below: Forest vegetation on Praslin

Butterfly Biology

Taxonomy

Butterflies and moths both belong to the insect order Lepidoptera. This insect order is one of the largest with more than 150,000 species found worldwide. There are approximately 18,000 butterfly species in 1900 genera (Shields 1989). These insects range in size from small Leaf-miner moths to the large Birdwing butterflies, with 3 mm and 300 mm wingspans respectively. Adult Lepidoptera are distinguished from other insects by a covering of broad, often brightly coloured, overlapping scales on the wings. Furthermore, the mouthparts of the adult are highly modified to form a coiled haustellum.

Butterflies can be readily distinguished from moths by the following characteristics:

1) Butterflies are generally day-flying, whereas moths are most active at night. Nonetheless, there are numerous day-flying moths as well;

2) Generally, butterflies settle with their wings folded above the body (except for the Pyrginae), and moths rest with their wings folded along the body or held flat;

3) Butterfly antennae end in a thickened club and moth antennae are either branched or thin and pointed.

Butterflies comprise three superfamilies: the Hesperioidea (skippers), Papilionoidea (true butterflies) and Hedyloidea. The Hesperioidea and Papilionoidea include approximately 3500 and 15,000 species respectively. The Hedyloidea are restricted to tropical America including Cuba and Trinidad, with about 35 species in a single genus, *Macrosoma*.

Both the Hesperioidea and Papilionoidea occur in Seychelles. The higher classification of the Seychelles butterflies is as follows:

Class: Insecta
Subclass: Pterygota
Division: Endopterygota
Order: Lepidoptera
  Superfamily: Hesperioidea
    Family: Hesperiidae
      Subfamily: Coeliadinae
      Subfamily: Pyrginae
      Subfamily: Hesperiinae
  Superfamily: Papilionoidea
    Family: Papilionidae
      Subfamily: Papilioninae
    Family: Pieridae
      Subfamily: Coliadinae
      Subfamily: Pierinae
    Family: Nymphalidae
      Subfamily: Heliconiinae
      Subfamily: Danainae
      Subfamily: Satyrinae
      Subfamily: Nymphalinae
    Family: Lycaenidae
      Subfamily: Theclinae
      Subfamily: Polyommatinae

Life cycle & morphology

This book primarily deals with the identification of the adult butterflies found in Seychelles. However, this section will aid identification of the early life-history stages, and used in conjunction with the descriptions of the eggs, larvae and pupae will allow the reader to identify the early stage family. Furthermore, identifying the plant species on which an early stage is found will often help narrow down which butterfly species it is.

All butterflies pass through four life-history stages, in a process called metamorphosis. The first stage is the egg (ovum), followed by the larva (caterpillar), pupa (chrysalis) and adult (imago). Early stage morphology for each butterfly family is characteristic and easily identifiable to that higher taxon.

Many butterflies spend more time in their early stages than in the adult stage. Yet less is known about these stages of a butterfly's life cycle. The early stages, with their limited mobility, are often vulnerable to predation and parasitism.

The egg

The egg is generally laid on a suitable food plant, although some butterflies scatter eggs whilst in flight (e.g. Satyrinae). Some species use only a few larval food plants (oligophagous) or even a single species (monophagous), whereas other butterflies use a wide range of food plants (polyphagous).

Eggs are 'glued' to the food plant by a sticky secretion which quickly hardens. The number of eggs laid varies from species to species. Most butterflies lay their eggs singly (e.g. Pieridae), but some lay eggs in small clusters (e.g. Heliconiinae).

Every egg has a small depression at the top 'pole' in which there is a tiny hole, the micropyle. Sperm enters the egg through the micropyle. Air and moisture also diffuse through the micropyle while the embryo is developing. The egg surface can be smooth or covered with a pattern of ribs and ridges.

The egg stage for most butterflies is brief, lasting approximately one to two weeks, although this can vary.

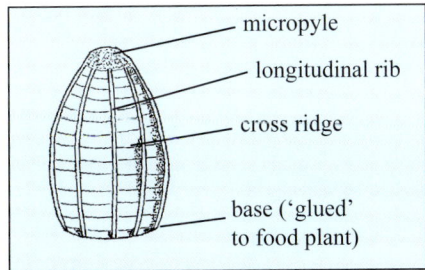

Above: Egg morphology
Below: Egg morphological variation

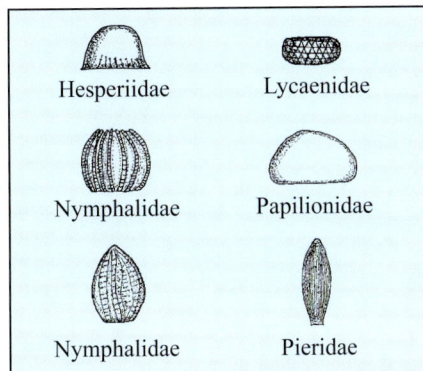

The larva

The larva is essentially the main feeding stage of a butterfly's life cycle. Many species devour the egg shell soon after hatching. The majority of butterfly larvae are herbivorous, however a number of lycaenid species can be carnivorous at certain times. As the larvae grow, they moult in stages known as instars. Generally, most butterfly larvae go through five instars, although there can be slight variations between species and even within a species.

The head is rounded, bears three pairs of simple eyes and the chewing mouthparts. The thorax comprises the first three segments behind the head. Each segment has a pair of jointed legs ending in a single claw. The abdomen is made up of 10 segments. Abdominal segments three to six and 10 each bear a pair of prolegs ending in tiny hooks called crochets. The prolegs support the elongated body of the larva. The larva breathes through spiracles.

Duration of the larval phase varies among species. In general, it lasts three to four weeks, but is often temperature dependent. During colder periods it can be considerably longer.

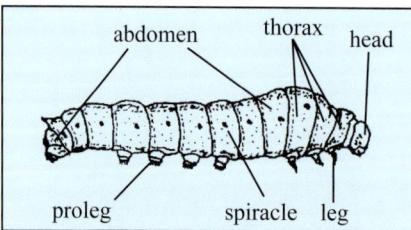

Hesperiidae	Papilionidae
Pieridae	Satyrinae
Danainae	Heliconiinae
Nymphalinae	Lycaenidae

Above: Larva morphological variation

The pupa

The pupal stage is a biological factory where the simple tissues of the larva are rearranged to form an adult butterfly. Most pupae are well camouflaged. Also, many pupae are dimorphic, being either various shades of green or grey/pink in colour.

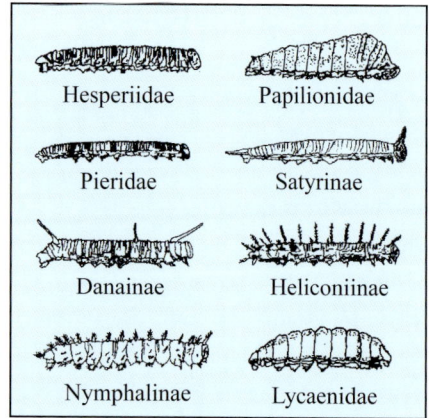

labial palpi
eye
foreleg
midleg
antenna
forewing
haustellum
abdominal segment
anus
cremaster

Above: Pupa morphology

The pupae of the nymphalid butterflies are suspended upside down from the cremasteral hooks located at the tail end.

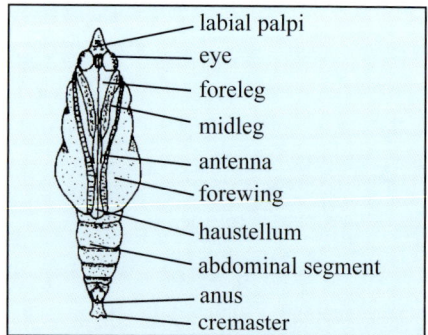

abdomen　thorax　head

proleg　spiracle　leg

Above: Larva morphology

These hooks are embedded in a mat of silk which is spun on the support structure by the larva before pupation takes place. The pupae of the Pieridae, Papilionidae and Lycaenidae are attached to the support structure by means of a cremaster as well as a girdle spun around the middle for added strength. Many hesperiids pupate in shelters (e.g. rolled grass leaves) constructed by the pupating larvae.

Pupal duration varies, but is usually three to four weeks for large species and 10 to 14 days for smaller butterflies.

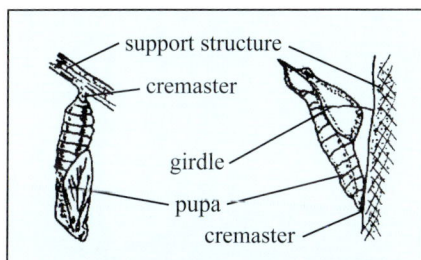

Above: Pupa attachment
Below: Pupa morphological variation

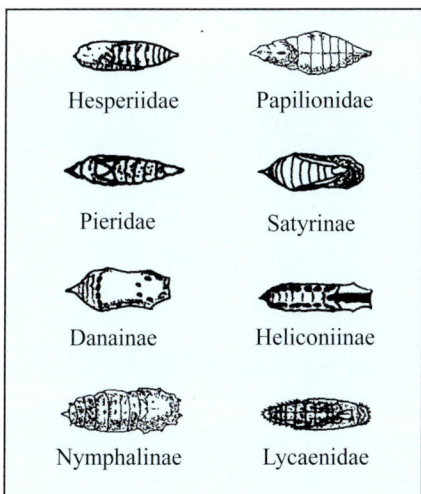

Hesperiidae	Papilionidae
Pieridae	Satyrinae
Danainae	Heliconiinae
Nymphalinae	Lycaenidae

The adult

The adult is essentially the breeding stage of a butterfly's life-cycle. Adults generally live for two to three weeks. As this book primarily deals with the identification of the adult butterfly, adult morphology is described in detail. Knowledge of imago morphology will assist in positively identifying its species.

The adult body is divided into the head, thorax (with legs and wings) and abdomen. The outer shell, known as the exoskeleton, is made of the polysaccharide called chitin. The muscles are attached to the exoskeleton. The structure of an adult butterfly can be broken down into the following main sections:

Head: A pair of long segmented antennae are found between two large compound eyes. The compound eye is made up of hundreds of facets called ommatidia. The mouthparts are greatly reduced compared with other insect orders. They consist of the maxillae which are modified to form the coiled haustellum or proboscis, and a pair of three-segmented labial palpi.

Thorax: The thorax consists of three sub-segments, the prothorax, mesothorax and metathorax. Each thoracic sub-segment carries a pair of jointed legs, while the meso and metathorax carry, in addition, a pair of wings each.

Legs: The jointed legs consist of five segments, the coxa, trochanter, femur, tibia and sub-segmented tarsus ending in a claw. In both sexes of the Hesperiidae, Papilionidae and Pieridae, the forelegs are fully developed and used for walking, but in the Lycaenidae and *Nymphalidae* are reduced in various degrees, especially in the males.

Wings: Butterflies, like most other insects, have two pairs of wings. The wings are membranous, and usually covered with pigmented scales, often of bright colours. These membranes are held rigid by veins. The membrane areas between the veins are known as cells. Furthermore, each general area of the wing has its own name. The veins, cells and general areas are used to describe the location of colour pattern features on the wings, and are useful for identifying a species.

Abdomen: The abdomen consists of 10 segments, and includes the respiratory, digestive and reproductive systems. The abdomen is generally much softer than the thorax. As butterflies and other insects have no lungs, oxygen enters through spiracles and is transported through the body via tiny tubes called tracheae. The stomach and midgut are found in the abdomen. The sex organs of a butterfly are proportionally large and are found in the last two (three in some species) segments of the abdomen. However, the gonads (i.e. the testes and ovaries) are found deeper inside the abdomen.

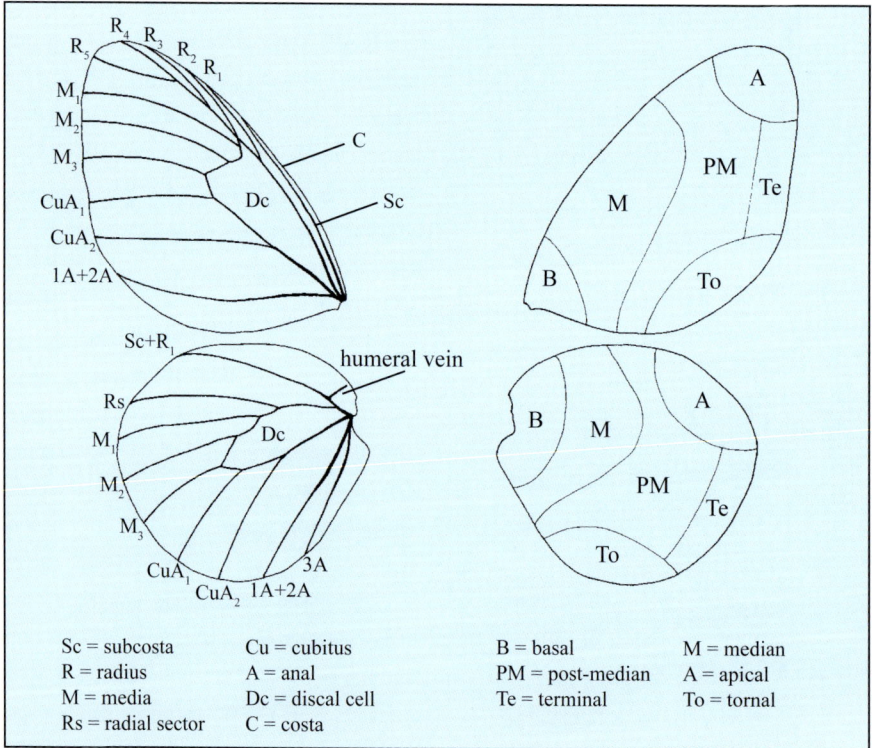

Sc = subcosta	Cu = cubitus	B = basal	M = median
R = radius	A = anal	PM = post-median	A = apical
M = media	Dc = discal cell	Te = terminal	To = tornal
Rs = radial sector	C = costa		

Above left: Wing venation (*Euploea mitra*); **Above right**: Wing general areas (*E. mitra*)

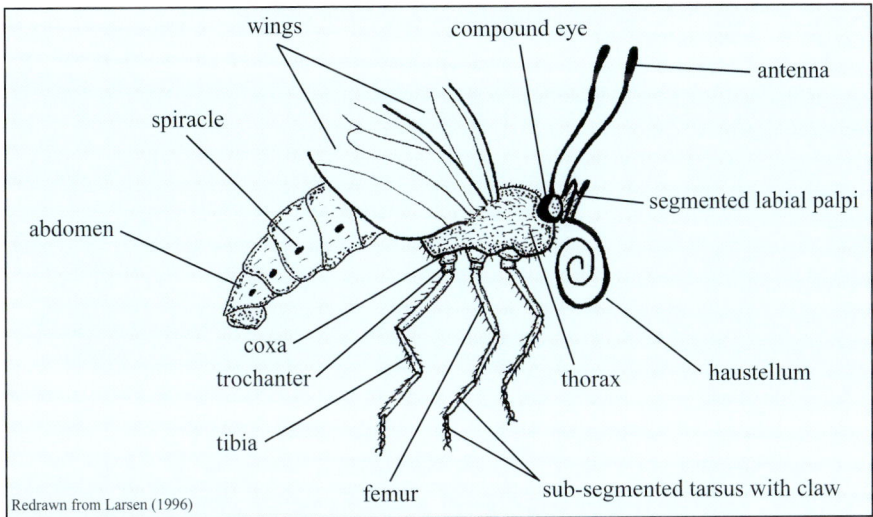

Labels on diagram: wings, compound eye, antenna, spiracle, segmented labial palpi, abdomen, coxa, trochanter, thorax, haustellum, tibia, femur, sub-segmented tarsus with claw

Redrawn from Larsen (1996)

Above: General structure of an adult butterfly

Labels on photograph: forewing, thorax, antenna, hindwing, segmented labial palpi, compound eye, abdomen, haustellum, tibia, femur, sub-segmented tarsus

South Africa

Above: General structure of an adult butterfly (*Vanessa cardui*)

Adult intraspecific variation

Four main types of phenotypic intraspecific variation are found within adult butterflies; sexual, seasonal, individual and geographical. All forms of intraspecific variation are illustrated in the *Species Accounts* section. Note that not all butterfly species exhibit intraspecific variation.

Sexual dimorphism

This refers to differences between the sexes of a particular species. In general, females are slightly larger than males, often with more rounded wings.

For most species the wing colour patterning of males and females is almost identical. In a number of species the only obvious difference between males and females is the presence of male secondary sexual characteristics. This is common in the Danainae butterflies. For example, *Euploea mitra* males have a sex-brand towards the trailing edge of each forewing, as well as hair-pencils at the end of the abdomen. Females lack these structures. The secondary sexual structures are used for producing and dispensing pheromones.

In some species the differences between male and female can be so great they appear to belong to different species. For example, many Lycaenidae butterflies have dull brown females and blue males. Generally with these species, the underside patterning is very similar in both sexes, and is more useful for identifying the particular species.

Seasonal variation

With seasonal variation, individuals that emerge during the dry season are different from those that hatch in the wet season. In the case of the Nymphalidae (e.g. *Junonia rhadama* and *Melanitis leda*), wet season forms show prominent marginal eyespot patterns (also known as ocelli), whereas these are absent in the dry season forms. Eyespots are displayed at rest and function principally in the deflection of vertebrate predator attacks. Wet season butterflies are also more active than dry season individuals.

With the Pieridae (e.g. *Colotis evanthe*) the wet season adults often have white undersides, while the dry season forms are more cryptically coloured. In the wetter tropics, species with seasonal forms elsewhere are monomorphic with the wet season phenotype, and the dry season form occurs as a rarity (Brakefield & Larsen 1984). This is the case with the two *Eurema* butterflies on Aldabra, where the dry season forms are rare or absent.

Individual polymorphism

This refers to phenotypic differences between individuals of the same species. Generally, the differences between individuals of the same population are not very pronounced. However, some butterflies have two or more different forms that occur at the same time in a particular population. This type of individual variation is known as polymorphism. For example, the sexually dimorphic *Hypolimnas misippus* has a monomorphic male, which is mostly black with large white spots on the wing upper surfaces, while there are four polymorphic female forms, which are mostly orange, with black and white forewing tips. Often intermediates between the four forms occur. The adaptive significance of this polymorphism is unknown.

Geographical variation

Variation may occur between geographically isolated populations of the same species. Populations that are isolated from each other may begin to evolve different characteristics which are geographically limited. If there are consistent differences between these separate populations, each is described as its own subspecies. In most cases the formation of a subspecies is aided by the presence of geographical barriers which limit gene flow between two populations.

Islands are geographically isolated and often have endemic subspecies of continental species. This is common in Seychelles, which has several endemic subspecies. For example, both *Eurema brigitta* and *Eurema floricola* are widespread butterflies throughout Africa as distinct subspecies. However, in Seychelles they both occur as the Malagasy endemic subspecies, *pulchella* and *aldabrensis* respectively. Furthermore, as Seychelles covers a wide expanse of the western Indian Ocean, the small skipper *Eagris sabadius* has two endemic subspecies in Seychelles (i.e. *maheta* occurs on the granitic islands and *aldabranus* on Aldabra Atoll).

Mimicry

There are two broad kinds of mimicry in adult butterflies.

In **Batesian mimicry**, a palatable species (the mimic) resembles an unpalatable one (the model), with the mimic benefitting by the protective resemblance. It is thought that a predator, after catching a number of individuals of the model, learns they are distasteful and avoids them. The predator will also avoid the mimic as they resemble the unpalatable butterflies. For this mimetic relationship to work, both mimic and model need to occur in the same area, with the model frequently attacked by a predator, so the predator learns which colour patterns to avoid. Flight behaviour of the model and mimic are also often different, with the model having a more lazy flight pattern. Furthermore, the non-flight posture of the model is more relaxed (e.g. drooping antennae) compared to the mimic.

In Seychelles, Batesian mimicry occurs between *Danaus chrysippus* and female *Hypolimnas misippus* f. *misippus*, and between *Danaus dorippus* and female *Hypolimnas misippus* f. *inaria*. *Danaus* species are the unpalatable models and *H. misippus* the palatable mimic. Male *H. misippus* are not mimetic (Gordon et al. 2010).

In **Müllerian mimicry**, an unpalatable butterfly resembles another unpalatable butterfly, with both benefitting. Müllerian mimicry does not occur in Seychelles.

MODEL | MIMIC

Danaus chrysippus | *Hypolimnas misippus* f. *misippus*

Danaus dorippus | *Hypolimnas misippus* f. *inaria*

Above: Mimetic resemblance between *D. chrysippus* and female *H. misippus* f. *misippus* (top) and, *D. dorippus* and female *H. misippus* f. *inaria* (bottom)

Below: Mimetic resemblance between *D. chrysippus* (left) and *H. misippus* female (right), showing different non-flight postures of the two species (South Africa)

Biogeography

Biogeography is the interpretation of species distributions (Whittaker & Fernández-Palacios 2007). It has important implications for the selection of biodiversity hotspots and conservation sites (Ladle & Whittaker 2011).

A total of 39 butterfly taxa have so far been recorded from the Seychelles islands. This is similar to the 40 (Williams 2007) and 32 (Martiré & Rochat 2008) butterfly taxa recorded from Mauritius and Réunion respectively. The much larger island of Madagascar has approximately 300 species (Lees et al. 2003), and the Comoros, which are the closest of the western Indian Ocean islands to continental Africa, has 93 taxa (Dall'Asta & Turlin 2009).

Of the Seychelles island groups, the Aldabra group has the highest richness with 25 taxa. The low-lying coral islands (i.e. Amirantes, Alphonse and Farquhar groups) have only eight taxa. Eighteen taxa are found in the Inner island group.

The butterfly fauna of the Inner islands and the Aldabra group can be considered fairly distinct from each other, even though both regions share several non-endemic taxa. The butterflies of the low-lying coral islands are all widely distributed cosmopolitan species that occur on both the Inner and Aldabra groups. Both the Inner and Aldabra groups have five endemic taxa each. *Borbo borbonica morella* is the only Seychelles endemic shared by both the Inner and Aldabra groups.

The butterfly fauna of the Inner islands is primarily derived from continental Africa, whereas the Aldabra group butterflies show a closer affinity to the Madagascan and Comoros fauna (Lawrence 2000a). All the endemic Seychelles subspecies have corresponding subspecies on either continental Africa, Madagascar, Réunion, Mauritius or Comoros.

The three endemic species (i.e. *Euploea mitra*, *Phalanta philiberti* and *Belenois aldabrensis*) are more difficult to link to surviving relatives and are of great biogeographical interest.

E. mitra, which is endemic to the granitic islands, is one of two Afrotropical species of a largely Indo-Oriental genus. The other Afrotropical species, *E. euphon*, is confined to the Mascarene islands, where it is represented by an endemic subspecies on each of the three islands. Although cladistic analysis suggest that both *E. mitra* and *E. euphon* are both ancient *Euploea* species (Ackery & Vane-Wright 1984), there is nothing to suggest that they are closely related.

Although now considered extinct, *P. philiberti* was endemic to the granitic islands. It appears to have been closely allied to the Indo-Oriental *P. alcippe* and the Madagascan *P. madagascariensis*.

B. aldabrensis, which is endemic to the Aldabra group, is quite different from other *Belenois* species (Bernardi 1954). It may be a remnant population of a formerly more widespread species.

Conservation

The Seychelles butterfly fauna is primarily threatened by habitat destruction (Gerlach 2008). This has resulted in a decrease in the numbers of several endemic butterflies and the extinction of others.

Approximately 46% of Seychelles is under some form of protection (Tingay 2010). Aldabra Atoll, which has the highest butterfly biodiversity in Seychelles, is protected as a World Heritage Site. The following taxa are endemic to Aldabra: *Eagris sabadius aldabranus*, *Eurema floricola aldabrensis*, *Colotis evanthe evanthides* and *Belenois aldabrensis*. On the granitic islands, the endemic taxa tend to be confined to the higher mountains of Mahé and Silhouette, where *Borbo borbonica morella* and *Eagris sabadius maheta* are regularly seen. Protecting the habitat of these endemic butterflies is important for their long-term survival.

Despite high levels of habitat destruction in tropical areas, very few threatened tropical butterflies are the focus of single-species conservation efforts (New 2014). Unfortunately, this is the case in Seychelles, as such an approach is often expensive.

Four endemic Seychelles butterfly taxa are IUCN Red-Listed. These are *Euploea mitra*, *Phalanta philiberti*, *Papilio phorbanta nana* and *E. sabadius aldabranus* (Gerlach 1997; Gerlach & Matyot 2006).

Although abundant in the late 19th and early 20th centuries, *E. mitra* is now rarely encountered. Recently, individuals have occasionally been seen on Silhouette and Mahé at La Réserve, Le Niol and Mission. Historically, it has also been recorded from St. Anne, just off the north coast of Mahé, and a single capture on La Digue in 1909. This species is Red-Listed as 'Endangered'. The ecological requirements of this butterfly and the causes of its decline are unknown.

Historically, *P. philiberti* was common in marshy areas on Mahé, Praslin and Silhouette. The last specimen was collected in 1953, with more individuals observed between 1956 and 1960 (Legrand 1965). This butterfly has not been seen since and is considered extinct (Bernardi 1996). The reason for its decline is not fully known, but habitat degradation, especially the draining of marshlands, is probably the main cause.

P. phorbanta occurs as two subspecies, with the nominate one confined to Réunion, where it is Red-Listed as 'Vulnerable' (Collins & Morris 1985; New & Collins 1991). The Seychelles subspecies *nana* is only known from a pair collected before 1880. Given that this butterfly is not easily overlooked and that it has not been seen in Seychelles for over 100 years, it is almost certainly extinct.

E. sabadius aldabranus is confined to Aldabra where it is known from only six specimens (one female and five males) collected in 1906. However, due to its small size and drab colours it could have been overlooked. It is currently Red-Listed as 'Data deficient'.

Species Accounts

Overview

A total of 36 species are covered in this book. Three species are represented by two subspecies each, bringing the total number of taxa covered here to 39. All taxa are listed under their scientific names. Each family is briefly introduced. Butterfly taxonomy follows Ackery et al. (1995), updated to reflect recent taxonomic changes. Floral taxonomy for larval food plants follows Robertson (1989) and Friedmann (2011).

A total of 2512 specimens were examined from the following seven collections:

1) Natural History Museum, London, United Kingdom.
2) University Museum of Zoology, Cambridge, United Kingdom.
3) Oxford University Museum of Natural History, United Kingdom.
4) Cousine Island Collection, Seychelles.
5) Seychelles National Museum, Victoria, Mahé.
6) Ditsong National Museum of Natural History, South Africa.
7) J.M. Lawrence private collection, South Africa.

Abbreviations:	
f.	form
w.s.f.	wet season form
d.s.f.	dry season form
FW	forewing
HW	hindwing
LFW	wingspan (mm)
UNF	forewing ventral surface
UNH	hindwing ventral surface
UPF	forewing dorsal surface
UPH	hindwing dorsal surface
♂	male
♀	female

Below: Aldabra Atoll World Heritage Site

D. Lawrence

Page layout

The information for each taxon is presented on its own page(s) using the format below. The easiest way to identify a taxon is either to page through the *Species Accounts* section or to use the *Plates* section at the end of the book.

1
2
9

Coeliades forestan forestan (Stoll, 1784)

3
Afrotropical region
Not threatened

Plate 1
Striped Policeman

4

Description
Similar to *Coeliades forestan arbogastes*, except the green scales are replaced by beige-coloured scales. Sexes similar. LFW: ♂ 40–50 mm; ♀ 50–58 mm.

Distribution
5
The Afrotropical Region including Comoros, Mauritius, Rodrigues, Réunion and Seychelles, where it has been recorded on the islands of Mahé, La Digue, Aride and Cousine.

Biology and conservation
Behaviour and biotope preferences similar to the subspecies *arbogastes*. Adults have been recorded throughout the year, but it is more commonly encountered during the wet NW Monsoon period. This butterfly occasionally migrates.

7

8

♀

beige scales   white band

South Africa   S. Woodhall

10

Larvae are polyphagous and feed on a wide range of food plants which include *Canavalia* species, *Terminalia catappa* and *Parinari curatellifolia*. Vinson (1938) considered this butterfly a minor pest on *Canavalia* species in Mauritius.

This subspecies is widespread and common across most of the Afrotropical region and under no threat.

11

Above top: *C. forestan forestan*
Above bottom: Male *C. forestan forestan* scent marking
Left: Seychelles distribution of *C. forestan forestan*

6

Granitic group

N

50 km

N

INDIAN OCEAN
200 km

30

Number key explanation

1   **Species name**: Scientific name of the taxon, and the name of the author (with date) who described it. If the author's name is in parentheses, it means that the taxon now belongs to a different genus from that in which it was originally placed when first described.

2   **Common name**: The name most widely used.

3   **Summary information**: Summary of the butterfly's distribution and conservation status.

4   **Description**: A description highlighting the important identification characteristics of the imago is given. Mean wingspan (mm) between forewing tips of a set specimen is provided.

5   **Distribution**: The Seychelles islands where a taxon has so far been recorded is listed here. In addition, a brief overview of the Afrotropical distribution of each non-endemic taxon is given.

6   **Map**: The Seychelles distribution of each taxon is illustrated, with its primary Seychelles distribution covered in detail in the inset map.

7   **Biology and conservation**: The biotope preference and general behaviour of each species is outlined in this section, including known adult phenology. All known Seychelles larval food plants are listed. In addition, the global conservation status of each butterfly and any specialised information such as taxonomic changes are dealt with here.

8   **Illustration**: Each taxon is illustrated with an oil painting. On each painting, significant identification characteristics are pointed out. Butterflies have been illustrated with the dorsal surface of the wing on the left, attached to the thorax, and the ventral surface on the right, separated from the thorax. In a few cases complete dorsal and ventral wing surfaces have been illustrated. All Seychelles phenotypic variations found within a particular taxon have been illustrated.

9   **Plate**: Each oil painting can also be found in a series of *Plates* at the end of the book. These *Plates* will help with differentiating similar looking taxa.

10   **Photograph**: Where no taxonomic differences occur, photographs of butterflies from areas outside Seychelles (mainly South Africa) have been used for certain taxa. This was done due to the limited availability of good quality photographs from Seychelles. The locality of non-Seychelles photographs are provided on the picture itself. Despite this limitation, the photographs nevertheless give an idea of the natural posture of a species. Unfortunately, photographs were not available for all taxa.

11   **Higher classification**: The family and subfamily to which a taxon belongs is colour-coded.

Hesperiidae

The Hesperiidae, commonly known as the skippers, form a cosmopolitan family with 3000 to 4000 species, currently distributed among approximately 570 genera (Warren et al. 2008). About 530 species occur in the Afrotropical region. Seven subfamilies are recognised (Warren et al. 2009), with the Coeliadinae, Pyrginae and Hesperiinae occurring in Seychelles.

Skippers are generally drab in colour, with broad wings and thick bodies. The head is broad. All three pairs of legs are fully developed for walking in both sexes. Adult flight is rapid and often appears erratic. At rest, the wings may be held vertically above the body, or horizontal and extended.

Five species occur in Seychelles. Three are represented by two subspecies each, making the total number of taxa eight. Other than these eight taxa, one other species, represented by two subspecies, has been historically associated with Seychelles and warrants discussion.

1) *Parnaro naso poutieri*. Identified from a worn specimen collected on Mahé, (Holland 1896), and possibly confused with *Borbo borbonica morella*. *P. naso poutieri* is endemic to Madagascar (Ackery et al. 1995) and is not considered part of the Seychelles butterfly fauna.

2) *Parnaro naso naso*. Apparently collected by J.A. de Gaye on Silhouette in the early 1900s. However, Fletcher (1910 page 296) states *"Mr. de Gaye's collection contains numerous examples of Lepidoptera collected by him in Mauritius. I am inclined to question the true origin of this particular specimen, and this species must be queried as a Seychelles butterfly, pending further evidence"*. Evans (1937) also lists a single female from Mahé (as *Parnaro marchalii marchalii*, an earlier synonym of *P. naso naso*). However, this taxon is endemic to Mauritius, and as no confirmed specimen has been collected in Seychelles, it is not considered part of the Seychelles fauna.

The Seychelles Hesperiidae subfamilies

Coeliadinae: Recently revised by Chiba (2009) it consists of 78 species. Approximately 19 species occur in the Afrotropical region. Larvae are highly polyphagous.

Pyrginae: A large subfamily consisting of about 150 genera. Approximately 150 species occur in the Afrotropical region. Larvae feed on Poaceae and other Monocotyledons.

Hesperiinae: A large family of 330 Afrotropical species, with 2500 species known worldwide. Larvae feed mainly on Monocotyledons, such as Poaceae and Costaceae.

South Africa

S. Woodhall

Above: *Pelopidas mathias* (see *Species Accounts* page 36)

Coeliades forestan arbogastes (Guenée, 1863)

Malagasy subregion

Not threatened

Plate 1

Striped Policeman

Description

Sexes similar. LFW: 47–53 mm. UPF light-brown in ground colour with green overlay in basal area. UPH brown with green to white overlay covering most of basal area. UNF light-brown in ground colour with white scales between 1A+2A and trailing edge of wing. UNH light-brown with characteristic well-developed white band in post-median region. The thorax is covered with green scales.

Distribution

Madagascar and Seychelles, where it has been recorded on Cosmoledo Atoll. Incorrectly listed as occurring on the granitic islands in previous published works (Lawrence 2010).

Biology and conservation

The Striped Policeman is a strong flyer

♂

green scales

white band

usually found along forest edges and clearings. On Madagascar, it occurs in unnatural grasslands, forest margins and the littoral zone (Lees et al. 2003), with adults recorded between July and February. Nothing is known about its early stages.

This butterfly appears to have a sporadic distribution within Seychelles, most likely represented by vagrants from Madagascar. This taxon has been erroneously listed as occurring on Réunion (Legrand 1965) and Mauritius (Williams 2007). Although unconfirmed, *arbogastes* may also occur on Mayotte in the Comoros (Turlin 1995).

Despite the sporadic occurrence of this butterfly in Seychelles, it is widespread on Madagascar (Viette 1956). This taxon is not threatened at present.

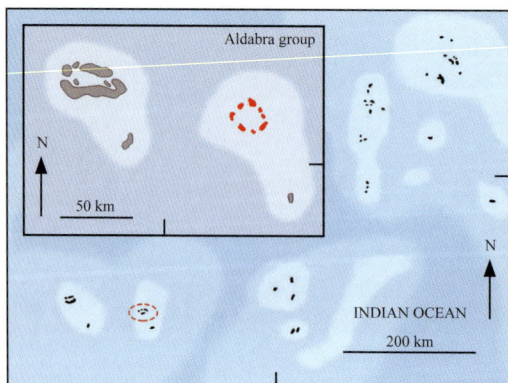

Aldabra group

N

50 km

N

INDIAN OCEAN

200 km

Above: *C. forestan arbogastes*

Left: Seychelles distribution of *C. forestan arbogastes*

Coeliades forestan forestan (Stoll, 1784)

Afrotropical region

Not threatened

Plate 1

Striped Policeman

Description

Similar to *Coeliades forestan arbogastes*, except the green scales are replaced by beige-coloured scales. Sexes similar. LFW: ♂ 40–50 mm. ♀ 50–58 mm.

Distribution

The entire Afrotropical region including the islands of Mauritius, Comoros, Réunion and Rodrigues. In Seychelles it has been recorded on the islands of Mahé, La Digue, Aride and Cousine.

Biology and conservation

Behaviour and biotope preferences similar to the subspecies *arbogastes*. Adults have been recorded throughout the year, but it is more commonly encountered during the wet NW Monsoon period. This butterfly occasionally migrates.

beige scales / white band

South Africa   S. Woodhall

The larvae are polyphagous and feed on a wide range of food plants which include *Canavalia* species, *Terminalia catappa* and *Parinari curatellifolia*. Vinson (1938) considered this butterfly to be a minor pest on *Canavalia* species in Mauritius. This subspecies is widespread and common across most of the Afrotropical region. The Striped Policeman is not threatened.

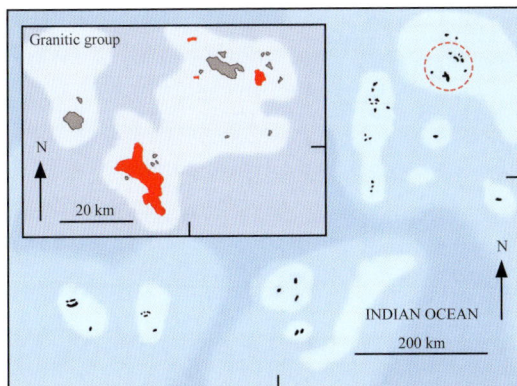

Granitic group

N

20 km

N

INDIAN OCEAN

200 km

Above top: *C. forestan forestan*
Above bottom: Male *C. forestan forestan* scent marking
Left: Seychelles distribution of *C. forestan forestan*

Eagris sabadius aldabranus Fryer, 1912

Seychelles endemic
Data deficient

Plate 1

Spotted Sylph

Description

Sexes dimorphic. ♂ LFW: 26–27 mm. UPF and UPH ochreous-brown, with poorly defined black markings. UNF ochreous-brown with dark-brown apex. UNH ochreous-brown with brown-black terminal and tornal areas. ♀ LFW: 33 mm. Similar to ♂ but with three adjacent hyaline spots in FW apex and three large well-defined hyaline spots in post-median areas. Similar to the subspecies *maheta* but is smaller, lighter in overall colour, with markings less distinct (Evans 1937). Ventral surfaces of the antennae lighter in colour (Fryer 1912).

Distribution

Endemic to Aldabra Atoll.

Biology and conservation

Fryer (1912) recorded a large number of specimens in *Pemphis acidula* bush. This butterfly is not widely collected, and is only known from five males and one worn female captured during the NW Monsoon period over 100 years ago. Early stages unknown.

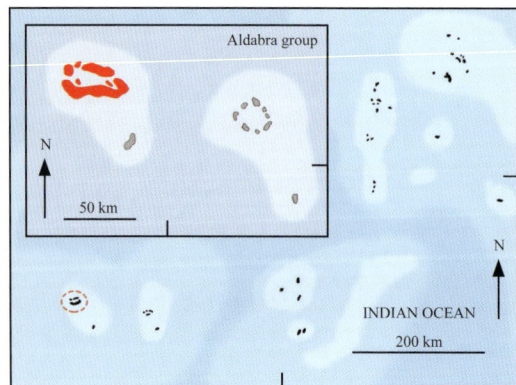

This taxon is listed as 'Data deficient' (Gerlach & Matyot 2006), with further research needed to be carried out on its biology and conservation status. Due to its small size and drab colour it may have been overlooked.

Above: *E. sabadius aldabranus*
Left: Seychelles distribution of *E. sabadius aldabranus*

Eagris sabadius maheta Evans, 1937

Seychelles endemic

Not threatened

Plate 1

Spotted Sylph

Description

Sexes dimorphic. ♂ LFW: 30–32 mm. ♀ LFW: 38–42 mm. Similar to *Eagris sabadius aldabranus* but larger and with prominent black markings. The defining characteristic of this taxon are the clear apical hyaline spots on the forewings of the ♂ (Evans 1937).

Distribution

Endemic to the granitic islands where it has been recorded on Mahé and Silhouette.

Biology and conservation

This butterfly flies along forest edges above 250 m a.s.l.. Occasionally it is seen lower. It is a strong flyer and frequently settles on exposed leaves with its wings held flat. Adults are most commonly encountered from November to April. Males are often seen patrolling along sunlit paths. The early stages of this taxon are unknown. Although it has only been recorded from two islands to date, it is common and under no immediate threat.

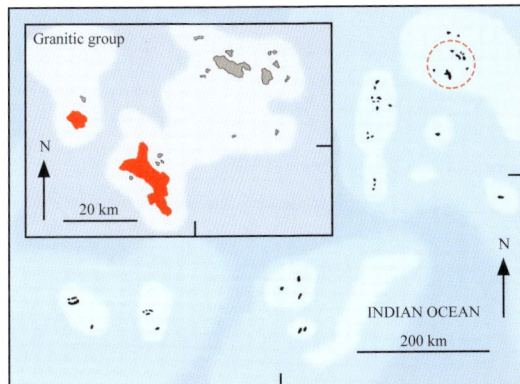

apical hyaline spots

prominent black markings

prominent hyaline spots

Above: *E. sabadius maheta*

Granitic group

N

20 km

INDIAN OCEAN

200 km

N

P. Mazzei

Above: *E. sabadius maheta*

Left: Seychelles distribution of *E. sabadius maheta*

Borbo borbonica morella (de Joannis, 1893)

Seychelles endemic
Not threatened

Plate 1
Olive Haired Swift

Description

Sexes similar. LFW: 34–37 mm. UPF dark-brown ground colour with variable hyaline spotting. Some specimens have a pale-brown spot between CuA_2 and 1A+2A, although it is often absent. UPH dark-brown ground colour. UNF red-brown ground colour, with post-median and trailing edge black-brown. UNH red-brown ground colour with a faint black-ringed white spot in post-median area, although in some individuals this is replaced with a small black spot.

small hyaline spots

pale-brown spot

overall red-brown

Distribution

Endemic to Seychelles where it has been recorded on Mahé, St. Anne, Silhouette, Cousine and Aldabra Atoll (see below).

Biology and conservation

This butterfly flies along forest edges generally 250 m a.s.l. and above. Males are territorial. Flight period is primarily from November to March. Larval food plant is the grass *Stenotaphrum dimidiatum*.

Initially described as a distinct species by de Joannis (1893). It was later classified as a subspecies of *Borbo borbonica* by Evans (1937) and has been treated as such since. As the genitalia of *morella* do not differ from the nominate subspecies (T. Larsen *pers. comm.*), the subspecific status of this taxon is maintained, despite the occasional appearance of the nominate subspecies in the granitic islands.

This butterfly may not be a permanent resident of Aldabra as there have been no records since 1907. As this butterfly is commonly encountered it is under no immediate threat.

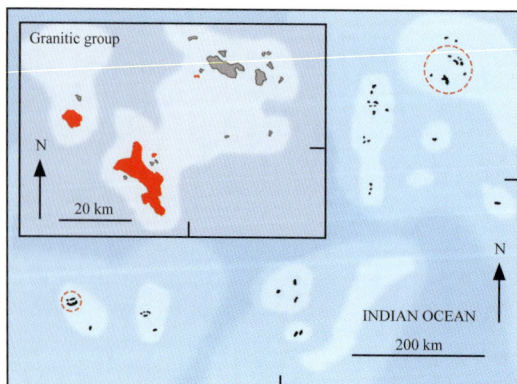

Above: *B. borbonica morella*
Left: Seychelles distribution of *B. borbonica morella*

33

Borbo borbonica borbonica (Boisduval, 1833)

Afrotropical region

Not threatened

Plate 1

Olive Haired Swift

Description

Sexes similar. LFW: ♂ 38–40 mm. ♀ 39–42 mm. Similar to *Borbo borbonica morella*, but larger, with more elongated FW, larger and more prominent hyaline spots, two or more brown-ringed white spots on UNH. UNF and UNH yellow-brown in colour.

elongated FW
prominent hyaline spots
overall yellow-brown

♂

Distribution

Most of the Afrotropical region including Madagascar, Mauritius, Réunion, Rodrigues, and Seychelles where it has been recorded on the islands of Mahé, Silhouette and Praslin.

♀

South Africa   S. Woodhall

Biology and conservation

This butterfly is an inhabitant of woodland, bush and open biotopes. Migration over the sea was observed on Mauritius during March 1996, with specimens arriving from the direction of Madagascar (Henning et al. 1997). It is relatively rare in Seychelles with no recent records. It is most likely that historical Seychelles records are of ephemeral populations. Recorded only during the month of April in Seychelles.

The larvae feed on *Pennisetum purpureum* and *Panicum maximum*. The Olive Haired Swift is common across the Afrotropical region and under no threat.

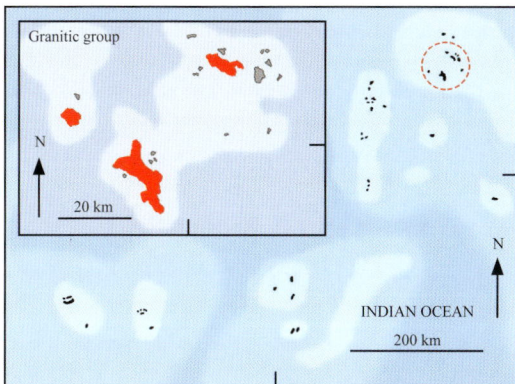

Granitic group

N

20 km

N

INDIAN OCEAN

200 km

Above top: *B. borbonica borbonica*
Above bottom: *B. borbonica borbonica* perching
Left: Seychelles distribution of *B. borbonica borbonica*

34

Borbo gemella (Mabille, 1884)

Afrotropical region

Not threatened

Plate 1

Twin Swift

Description

Sexes similar. LFW: ♂ 29–30 mm. ♀ 31 mm. Seychelles specimens are smaller than continental individuals (Berger 1962; Lawrence 2009a). UPF dark-brown ground colour with variable hyaline spotting. Poorly defined spot between CuA_2 and 1A+2A. UPH and UNF dark-brown ground colour. UNH dark-brown with faint white scaling giving overall grey-brown appearance. Two or more unringed white spots. Additional image: page 124.

poorly defined spot

overall grey-brown

♂

South Africa   S. Woodhall

Distribution

Afrotropical region including parts of the Middle East, Madagascar, Comoros and Seychelles, where it has been recorded on the islands of Mahé, Silhouette, Praslin, La Digue, Aride, Cousine, Cousin, Coëtivy, Alphonse, Platte and Aldabra Atoll.

Biology and conservation

This butterfly is found along forest edge biotopes and is most active in the early morning (Lawrence 2004a). Seen all year round, but most common from November to January. Larval food plants are grasses such as *Stenotaphrum dimidiatum* and *Zea* species. This butterfly is widespread, often encountered and not threatened.

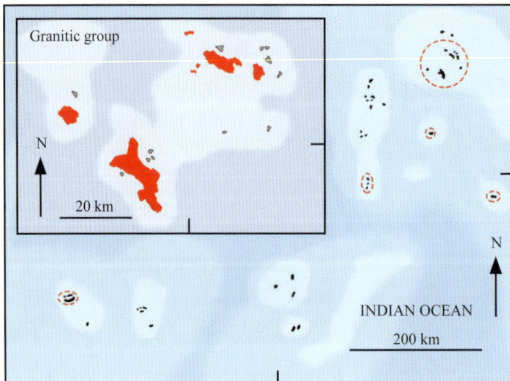

Granitic group

N

20 km

N

INDIAN OCEAN

200 km

Above top: *B. gemella*

Above bottom: *B. gemella* mating pair

Left: Seychelles distribution of *B. gemella*

Pelopidas mathias (Fabricius, 1798)

Afrotropical region
Not threatened

Plate 1
Black Branded Swift

Description

Sexes similar. LFW: ♂ 33–36 mm. ♀ 36–38 mm. UPF dark-brown ground colour with hyaline spots. Male with a black sex-brand, which is replaced with a well-defined hyaline spot in the female. UPH brown with extensive covering of green-yellow scales. UNF brown with faint white scaling. UNH brown with widespread faint white scaling and numerous post-median white spots. Additional image: page 28.

♂

black sex-brand

overall grey-brown

♀

well-defined spot

overall grey-brown

Distribution

Widespread throughout the Afrotropical and Oriental regions including the Indian Ocean islands. In Seychelles only recorded from Astove.

Biology and conservation

So far it has only been recorded on Astove

on 8 March 1967 (Lionnet 1970). As there are no other Seychelles records it is possible this species is not resident and may represent a straggler or an ephemeral population. Larsen (1996, 2005) considers this species migratory. The Black Branded Swift is polyphagous on various grasses such as *Zea*, *Panicum*, *Hyparrhenia* and *Oryza* species, all of which occur in Seychelles (Robertson 1989). It is common and widespread outside Seychelles and under no immediate threat.

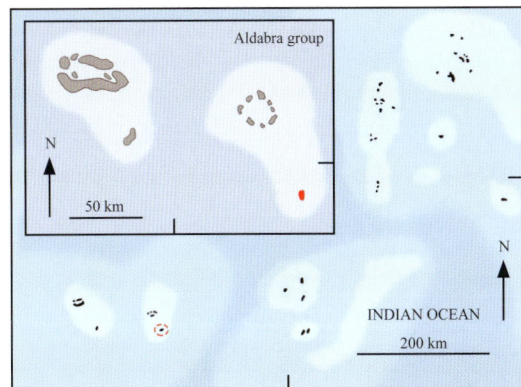

Aldabra group

N

50 km

N

INDIAN OCEAN

200 km

Above: *P. mathias*

Left: Seychelles distribution of *P. mathias*

Papilionidae

The Papilionidae, commonly known as the swallowtail butterflies, form a cosmopolitan family with approximately 570 species (Collins & Morris 1985). Of the three recognised extant subfamilies (i.e. Papilioninae, Parnassiinae and Baroniinae), the Papilioninae are the most diverse. The Baroniinae are represented by one species restricted to SW Mexico. The Parnassiinae comprise approximately 50 species and are confined to the Holarctic region. A fourth subfamily, the Praepapilionae, which are now extinct, consisted of two species from North America. The Praepapilionae were closely related to the Baroniinae.

Swallowtails are large butterflies that are generally brightly coloured. The hindwings are often 'tailed', the antennae are short and the maxillary palpi reduced. The forelegs are fully functional in both sexes. The adults are usually strong, rapid flyers and some species are thought to be migratory. Males are often territorial. They are regularly seen feeding from flowers, especially in the morning, with the wings constantly vibrating and held vertically above the body.

The swallowtail butterflies are poorly represented in Seychelles, with only two species attributed to these islands. One, the endemic *Papilio phorbanta nana* is now extinct, and the other, *Papilio dardanus* recorded as a single sighting only.

P. phorbanta nana belongs to the metallic-blue-banded clade of swallowtails known as the *nireus* group. These butterflies are characterised by the large metallic-blue bands across the upper surfaces of the fore and hindwings of the males. In the females these blue bands are often reduced or even absent.

This group of butterflies is well represented in the Malagasy subregion where it has speciated into five endemic species.

Distribution of the *nireus* swallowtail group of butterflies in the Malagasy subregion:

1) *Papilio manlius* – Mauritius

2) *Papilio phorbanta* (two subspecies)
    phorbanta – Réunion
    nana – Seychelles

3) *Papilio oribazus* – Madagascar

4) *Papilio epiphorbas* (three subspecies)
    epiphorbas – Madagascar
    guyonnaudi – Comoros
    praedicta – Comoros

5) *Papilio aristophontes* – Comoros

Papilio phorbanta nana Oberthür, 1880

Seychelles endemic
Extinct

Plate 2
Papillon La Pâture

Description

Sexes dimorphic. ♂ LFW: 65 mm. UPF black with large blue spots in discal cell and post-median area. UPH black with broad blue median band in basal and post-median areas. UNF ground colour black-brown. UNH ground colour similar to UNF but with prominent row of white submarginal spots. ♀ LFW: 74 mm. UPF, UPH, UNF and UNH red-brown ground colour with grey-white submarginal spots and bands.

♂

short tail
blue band

♀

short tail
grey-white band

Distribution

Endemic to Seychelles. Assumed to be collected on Mahé. Nominate subspecies endemic to Réunion.

Biology and conservation

Known only from a single male and a single female allegedly collected from Seychelles before 1880. They could represent wind-blown vagrants from Réunion (Paulian & Viette 1968). Alternatively, this taxon could have been intentionally (Hancock 1983) or unintentionally introduced via the establishment of *Citrus* trees from Réunion in the 1700s and 1800s (Legrand 1959). Larvae of the nominate subspecies feed on Rutaceae (Martiré & Rochat 2008). It has not been seen for over 100 years and is considered 'Extinct'.

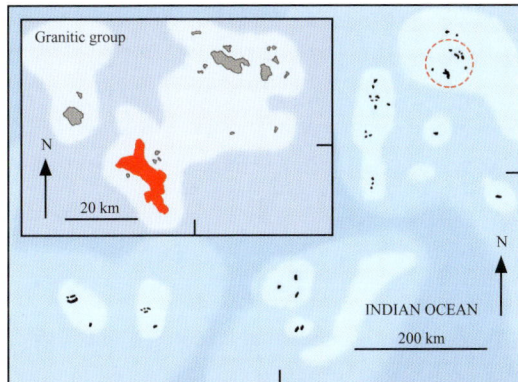

Granitic group
N
20 km
N
INDIAN OCEAN
200 km

Above: *P. phorbanta nana*
Left: Assumed distribution of *P. phorbanta nana in* Seychelles

Papilio dardanus Brown, 1776

Afrotropical region
Not threatened

Plate 2

Mocker Swallowtail

Description

LFW: 84–90 mm. UPF sulphur-yellow with marginal areas black. Small sulphur-yellow spot in apical area. UPH sulphur-yellow with post-median areas black, tailed. UNF similar to UPF but with brown marginal areas. UNH basal areas sulphur-yellow, median and post-median areas with black to dark-brown band, terminal area yellow-brown, veins darkened.

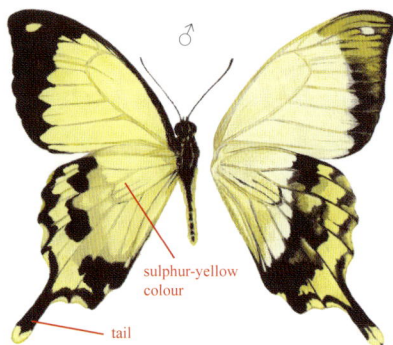

♂

sulphur-yellow colour

tail

Distribution

Afrotropical region including Madagascar, Comoros and Seychelles, where a single sighting has been recorded on Aldabra Atoll.

Biology and conservation

The Mocker Swallowtail is widespread across the Afrotropical region, with about 13 described subspecies. Sexual dimorphism is generally marked in the continental subspecies, where females are polymorphic and normally without tails. However, females of the subspecies *antinorii* (Ethiopia and Somalia), *meriones* (Madagascar) and *humbloti* (Comoros) resemble the males.

In Seychelles, this butterfly is known from a single sighting on Aldabra on 22 November 1959 (Legrand 1965) of what was most likely a vagrant from either Madagascar or the Comoros. This species is generally associated with a forest biotope and Aldabra is considered unsuitable for it. Elsewhere, larvae feed on Rutaceae (e.g. *Citrus* species), which are absent from Aldabra. This butterfly is under no immediate threat.

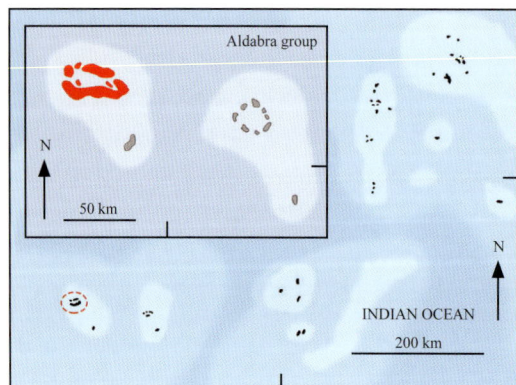

Aldabra group

N

50 km

N

INDIAN OCEAN

200 km

Above: *P. dardanus*
Left: Seychelles distribution of *P. dardanus*

Pieridae

The Pieridae, commonly known as the whites and sulphurs, form a cosmopolitan family with approximately 1100 species, currently arranged in 83 genera (Braby et al. 2006). Four subfamilies are recognised worldwide, with the Coliadinae and Pierinae occurring in Seychelles.

Generally, these butterflies are medium-sized, are white or yellow with black-margined wings. Some species have red-coloured forewing tips. The hindwings are never tailed. The sexes are often dimorphic. In many species that fly throughout the year, wet season generations are different from dry season ones. Wet season forms are larger with heavier black markings, while the dark markings of the dry season broods are greatly reduced, and often the undersides are pinkish-brown in colour.

The phylogenetic position of the pierids in relation to the other butterfly families is uncertain. They are considered to be the sister family to either the Papilionidae or the Nymphalidae (Wahlberg et al. 2005). However, as in the Papilionidae, they have six normal functional legs in both sexes, and the pupa is similarly attached by the cremaster and a silken girdle.

Flight is rapid and often close to the ground. They regularly congregate on moist sandy areas where they probe for moisture with their haustellum (i.e. mud-puddling). Many pierid species migrate.

Six species have so far been recorded from Seychelles. Two other species have been erroneously associated with Seychelles:

1) *Eurema desjardinsii* (Fryer 1912; Holland 1896) which has a subspecies that occurs on Comoros and Madagascar, and another subspecies which occurs in Africa.

2) *Colotis etrida* (Fryer 1912) which is an Oriental species and was most likely confused with the similar looking *Colotis evanthe evanthides*.

The Seychelles Pieridae subfamilies

Coliadinae: There are some 240 members of this subfamily worldwide. They are poorly represented in the Afrotropical region with just 15 species. The larval food plants are mainly the Fabaceae, although the *Eurema* feed on several other plant families.

Pierinae: This is a cosmopolitan subfamily with about 700 members. Approximately 170 species occur in the Afrotropical region. Primary larval food plants are the Capparaceae and the Loranthaceae.

Catopsilia florella (Fabricius, 1775)

Afrotropical region
Not threatened

Plate 3
African Migrant

Description

Sexes dimorphic. ♂ LFW: 54–60 mm. UPF green-white ground colour. Margins black-bordered. UPH pale-green-white with red-brown margins. UNF and UNH pale-green-white. Has a hair-pencil on ventral surface of FW between CuA_2 and 1A+2A. ♀ LFW: 56–66 mm. ♀ polymorphic. One form is similar to the ♂, but with the pale-green-white replaced with pale-yellow. The other ♀ form resembles the ♂.

Distribution

The Afrotropical region including the Indian Ocean islands. Only recorded on Mahé in Seychelles.

Biology and conservation

Known only from a single specimen collected by R.P. Philibert (de Joannis 1894), which

green-white colour

yellow colour

probably represents a wind-blown vagrant. As its common name suggests this butterfly regularly migrates.

Elsewhere the larvae are polyphagous on Caesalpiniaceae, Malvaceae, Mimosaceae and Papillionaceae plants, of which there are many species in Seychelles. This butterfly is not considered a Seychelles resident species at the moment. The African Migrant is commonly encountered, widespread and not threatened.

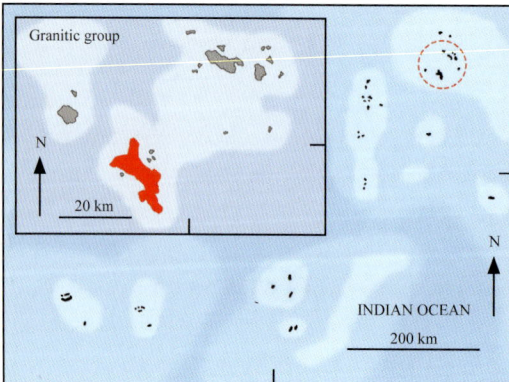

Granitic group

N

20 km

N

INDIAN OCEAN

200 km

Above: *C. florella*

Left: Seychelles distribution of *C. florella*

41

Top: Male *C. florella*; **Inset**: Hair-pencil on male *C. florella* forewing ventral surface
Bottom: Female *C. florella*

Eurema brigitta pulchella (Boisduval, 1833)

Malagasy subregion
Not threatened

Plate 3

Broad Bordered Grass Yellow

Description

Sexes dimorphic. ♂ LFW: 34–41 mm. UPF and UPH yellow ground colour with broad black borders. UNF and UNH generally yellow ground colour with variable black dusting. ♀ LFW: 30–39 mm. Similar to male, but ground colour paler, FW costal border narrower, and no black border on UPH. All wing surfaces more extensively black dusted. Additional image: Plate 3.

♂

broad black margins

w.s.f.

♀

extensive black dusting

w.s.f.

Distribution

This taxon occurs on Madagascar, Mauritius, Comoros and Seychelles, where it has been recorded on Aldabra. A single specimen has been collected on Réunion at La Montagne in 1955 (Viette 1957)

Biology and conservation

The wet season form of the Broad Bordered Grass Yellow is common on Aldabra. The flight is slow and close to the ground. They are often seen feeding on flowers. The dry season form (Plate 3) is exceptionally rare on Aldabra. This butterfly is generally on the wing during the NW Monsoon period.

Larval food plants include *Tephrosia* species (Williams 2007) of which there are numerous species in Seychelles (Robertson 1989). This subspecies is common and widespread across the Malagasy subregion and under no immediate threat.

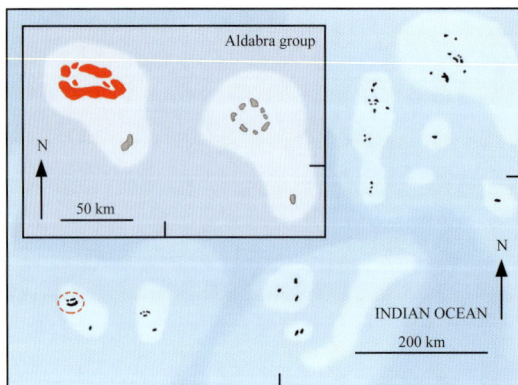

Aldabra group

N

50 km

N

INDIAN OCEAN

200 km

Above: *E. brigitta pulchella*
Left: Seychelles distribution of *E. brigitta pulchella*

43

Eurema floricola aldabrensis (Bernardi, 1968)

Seychelles endemic
Not threatened

Plate 3

Malagasy Grass Yellow

Description

Sexes dimorphic. ♂ LFW: 31–32 mm. UPF and UPH lime-yellow ground colour with a broad black border on the FW. UNF and UNH lime-yellow, slightly paler than dorsal surfaces and with well-defined black-ringed white spots. ♀ LFW: 30–33 mm. Similar to ♂, but ground colour paler yellow. FW costal border barely traceable below R$_2$. Apical patch present on UNF.

Distribution

This taxon is endemic to Seychelles where it has been recorded on Aldabra and Astove.

Biology and conservation

The Malagasy Grass Yellow is another frequently encountered butterfly on Aldabra. Similar to the previous species, its flight is slow and close to the ground. Both sexes are attracted to flowers. Male butterflies mud-puddle. This butterfly flies throughout the year, but it is more abundant in January, April and May.

Only the wet season form has been recorded in Seychelles so far. The nominate subspecies, which occurs on Madagascar, has well-defined dry and wet season forms (Yata 1994).

The early stages of this taxon are unknown. Although it has a narrow distribution, it is common, and at this stage under no immediate threat.

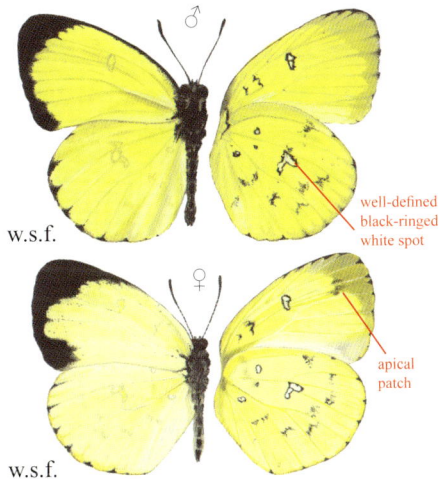

w.s.f.

well-defined black-ringed white spot

apical patch

w.s.f.

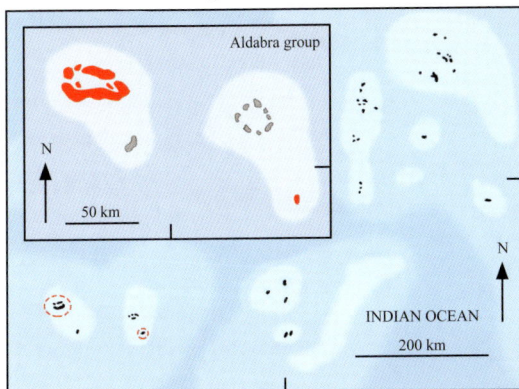

Aldabra group

N

50 km

INDIAN OCEAN

200 km

Above: *E. floricola aldabrensis*
Left: Seychelles distribution of *E. floricola aldabrensis*

Coliadinae

Belenois grandidieri (Mabille, 1878)

Malagasy subregion
Not threatened

Plate 4
Malagasy White

Description

Sexes dimorphic. ♂ LFW: 52 mm. UPF and UPH white ground colour. FW apex black with white patches. Discal cell apex margin black-bordered. Patches of yellow in basal areas of both FW and HW. UNF and UNH similar to dorsal wing surfaces. ♀ LFW: 55 mm. Similar to ♂, but with heavier black markings.

Distribution

Madagascar and Seychelles, where it has been recorded on Astove.

Biology and conservation

So far this butterfly has only been recorded on Astove on 8 March 1967 (Lionnet 1970). It is only known in Seychelles from a single male and a single female. As there are no other Seychelles records it is possible this species is not resident and may represent an ephemeral population. Alternatively, Gerlach & Matyot (2006) suggest that this species could be a dry season form of *Belenois aldabrensis*. Further work on the Malagasy *Belenois* species is required to clarify their taxonomy.

The early stages of this butterfly are unknown. This species is widespread across Madagascar, common and not threatened at present.

Above: *B. grandidieri*
Left: Seychelles distribution of *B. grandidieri*

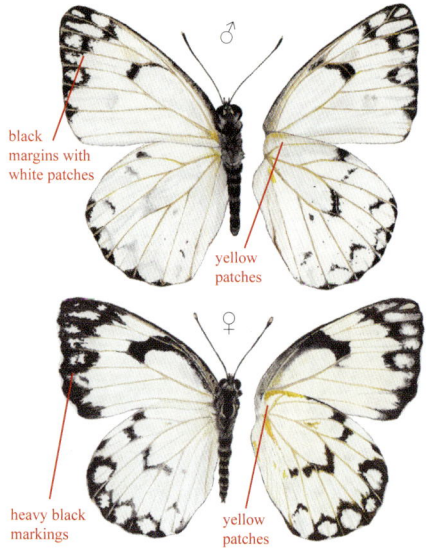

♂

black margins with white patches

yellow patches

♀

heavy black markings

yellow patches

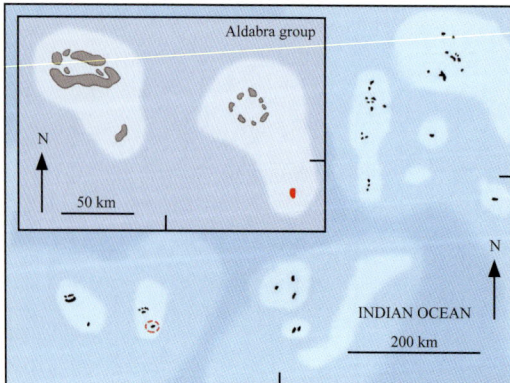

Aldabra group

N

50 km

INDIAN OCEAN

N

200 km

Belenois aldabrensis (Holland, 1896)

Seychelles endemic
Not threatened

Plate 4

Aldabra White

Description

Sexes dimorphic. ♂ LFW: 35–36 mm. UPF and UPH white ground colour with thin black border on the FW apex and margin. UNF white ground colour with pale-yellow apex and wing margin. UNH pale-yellow ground colour. ♀ LFW: 35–37 mm. Similar to ♂, but with heavier black and darker yellow markings. The apical area of the UPF is black-bordered, enclosing five pale-yellow-white spots. UNF with a post-median series of three grey-black spots.

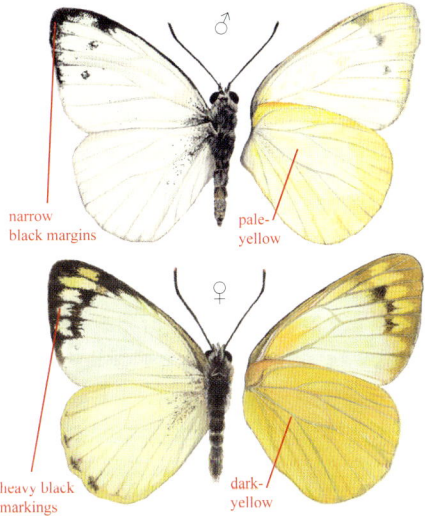

narrow black margins

pale-yellow

heavy black markings

dark-yellow

Distribution

This species is endemic to Seychelles where it has been recorded on Aldabra, Assumption and Astove.

Biology and conservation

The Aldabra White butterfly is frequently encountered on the Aldabra group islands. Its flight is fast, and it appears to be most active between 10h00 and 14h00 (Betts 2000). This butterfly flies throughout the year, but is more abundant from September to March

The early stages of this taxon are unrecorded, but larvae have been seen on *Morinda citrifolia* on Aldabra (Legrand 1965). Although this butterfly has a narrow distribution, it is common in the Aldabra group and is at this stage under no immediate threat.

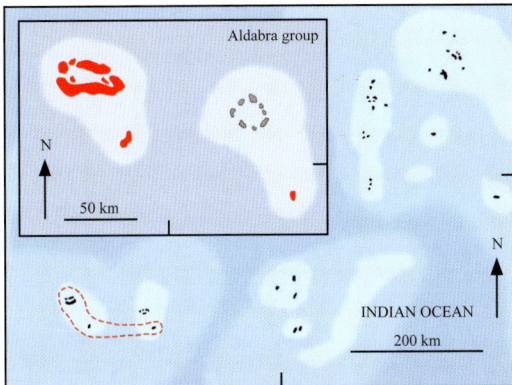

Aldabra group

N

50 km

INDIAN OCEAN

200 km

N

Above: *B. aldabrensis*

Left: Seychelles distribution of *B. aldabrensis*

46

Colotis evanthe evanthides (Holland, 1896)

Seychelles endemic
Not threatened

Plate 4

Aldabra Red Tip

Description

Sexes dimorphic. ♂ LFW: 28–36 mm. UPF and UPH white with variable grey scaling in basal and median areas. UPF apex with large black-bordered orange-red tip. UNF white, with yellow-orange apex. Black spot in discal cell. UNH yellow-orange ground colour. ♀ LFW: 29–38 mm. UPF with large orange apex. Black spot in discal cell. UPH white-orange with extensive basal grey scaling. UNF and UNH ground colour white-orange with black scaling. Black spot in UNF discal cell. Additional image: Plate 3.

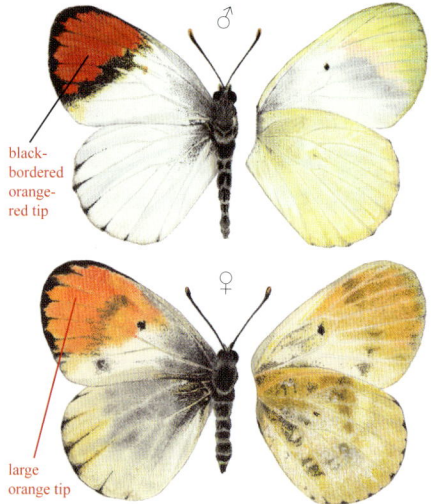

black-bordered orange-red tip

large orange tip

Distribution

Endemic to Seychelles where it occurs on Aldabra, Assumption and Astove.

Biology and conservation

The Aldabra Red Tip is often encountered on the Aldabra group islands. It is generally on the wing during the NW Monsoon period, and is most commonly seen between January and April. The early stages are unknown.

This butterfly was originally classified as the Seychelles endemic species *Colotis evanthides*. However, *C. evanthides* is so similar in morphology and DNA to *C. evanthe* from Comoros it is now considered a subspecies of the latter (Nazari 2011). Despite the narrow distribution of this taxon it is under no immediate threat.

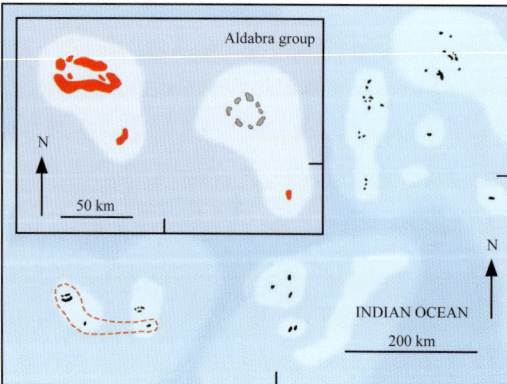

Aldabra group

N

50 km

N

INDIAN OCEAN

200 km

Above: *C. evanthe evanthides*
Left: Seychelles distribution of *C. evanthe evanthides*

Nymphalidae

The Nymphalidae, broadly known as the brush-footed butterflies, form a cosmopolitan family with approximately 7200 species. Some 1460 species occur in the Afrotropical region.

The phylogeny of this family has been frequently disputed, with the subfamily classification varying depending on the author. Twelve (Wahlberg et al. 2009) to 14 (Freitas & Brown 2004) subfamilies are recognised, with the Danainae, Heliconiinae, Satyrinae and Nymphalinae occurring on the Seychelles islands.

Generally these butterflies are medium to large in size. They vary greatly in colour pattern, behaviour, habitat choice and early stages. The sexes are generally dimorphic.

However, one characteristic that is shared by all nymphalid species is that the forelegs are highly reduced in both sexes. Thus, they only have four functional walking legs. The purpose of the reduced forelegs is unknown.

In many species that fly throughout the year, wet season generations are different from dry season ones. Wet season forms show prominent marginal eyespot patterns (Brakefield & Larsen 1984).

The Nymphalidae are the largest butterfly family in Seychelles with 17 species recorded so far. In addition to these 17 species there have been two unconfirmed sightings of what could be a *Neptis* species on Aride in May 1998 (Bowler & Hunter 1999) and March 2004 (Gerlach & Matyot 2006).

The Seychelles Nymphalidae subfamilies

Danainae: Revised by Ackery & Vane-Wright (1984), this subfamily consists of about 500 species worldwide. They are poorly represented in the Afrotropical region with approximately 24 species.

Heliconiinae: This subfamily consists of approximately 600 species worldwide. In Africa about 235 species are recognised. Larval food plants are mainly the Passifloraceae and the Urticaceae, although other plant families are often utilised.

Satyrinae: Found worldwide with roughly 1500 members. There are about 330 African representatives. Larval food plants are generally Monocotyledons, usually Poaceae, Arecaceae and Marantaceae.

Nymphalinae: A cosmopolitan subfamily with roughly 500 species. There are about 70 species in the Afrotropical region. Larval food plants include the Acanthaceae and the Urticaceae.

Danaus chrysippus orientis (Aurivillius, 1909)

Afrotropical region
Not threatened

Plate 5

Plain Tiger

Description

Sexes similar. LFW: ♂ 50–70 mm. ♀ 50–75 mm. UPF orange-brown from CuA_2 to trailing edge of wing. Discal cell red-brown. The apex is black with numerous white spots and patches. UPH orange-brown with black and white spotted margins. Three black spots where Rs, M_1 and M_2 join discal cell. ♂ has black sex-brand below discal cell, which ♀ lacks. UPH and UNF similar in colour pattern to the dorsal wing surfaces. The thorax has numerous white spots on the lateral surfaces. Additional images: Plate 5, page 21 and 94.

black and white apex

black spots

sex-brand

Distribution

The Afrotropical region including the Indian Ocean Islands. In Seychelles it has been recorded on the islands of Mahé, St. Anne, Silhouette, Praslin, La Digue, Aride, Denis Island, Cosmoledo, Aldabra and Astove.

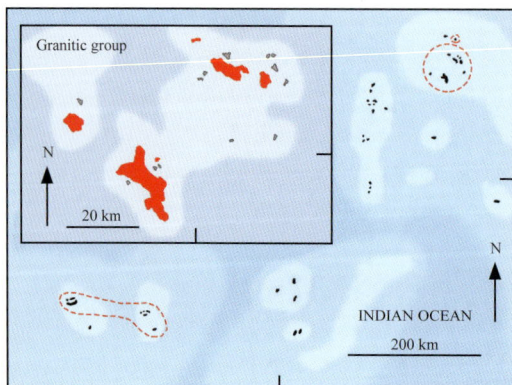

South Africa   S. Woodhall

Biology and conservation

This butterfly has a lazy flight pattern. It historically inhabited open savannah biomes, but has also colonised disturbed habitats in forest areas. It is migratory and flies throughout the year. Larvae feed on *Calotropis gigantea* in Seychelles. It is widespread and under no threat.

Granitic group

N

20 km

N

INDIAN OCEAN

200 km

Above top: *D. chrysippus orientis*
Above bottom: Freshly emerged *D. chrysippus orientis*
Left: Seychelles distribution of *D. chrysippus orientis*

Danaus dorippus dorippus (Linnaeus, 1758)

Afrotropical & Oriental regions
Not threatened

Plate 5
Dorippus Tiger

Description

Sexes similar. LFW: ♂ 50–70 mm. ♀ 50–75 mm. UPF orange-brown with black and white spotted leading edge, apex and terminal margins. UPH orange-brown with black margin. Three black spots where Rs, M_1 and M_2 join discal cell. ♂ has black sex-brand below discal cell, which ♀ lacks. UPH and UNF similar in colour pattern to the dorsal wing surfaces. Thorax has numerous white spots on the lateral surfaces. Additional images: Plate 5 and page 94.

Distribution

East Africa, Arabia, Iran, Pakistan and India. In Seychelles it has been recorded on the islands of Aldabra and Assumption.

Biology and conservation

This species has a lazy flight pattern. It is a

orange-brown apex
black spots
sex-brand
Kenya

butterfly of open biotopes. Both sexes feed on flowers. The early stages are unknown.

This butterfly was originally classified as a form of *D. chrysippus*. Interestingly, in East Africa *D. chrysippus* and *D. dorippus* interbreed and produce viable offspring (Smith et al. 2005). This taxon is widespread and under no threat.

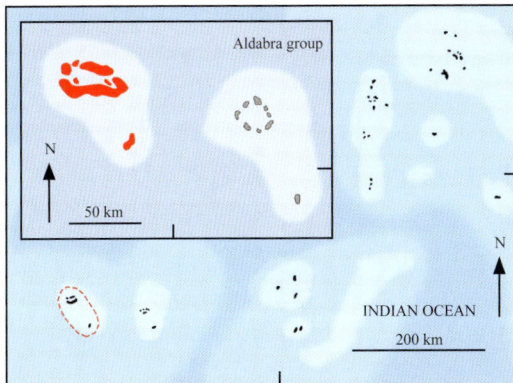

Above top: *D. dorippus dorippus*
Above bottom: *D. dorippus dorippus* feeding
Left: Seychelles distribution of *D. dorippus dorippus*

Aldabra group
N
50 km
N
INDIAN OCEAN
200 km

Amauris niavius dominicanus (Trimen, 1879)

Afrotropical region
Not threatened

Plate 5
Friar

Description

Sexes similar. LFW: ♂ 80–85 mm. ♀ 78–82 mm. UPF black with large white patches towards apex and trailing edge of the wing. UPH white with black margins and dark-brown tornal area. UNF and UNH similar in colour pattern to the dorsal wing surfaces. Thorax has numerous white spots.

black with large white patches

Distribution

Eastern and southern Africa. In Seychelles it has been recorded on the island of Mahé.

South Africa   S. Woodhall

Biology and conservation

So far the Friar is known only from a single capture on Mahé at Bel-Air by M. Mason in 1953 (Legrand 1965). It may represent an ephemeral population or a wind-blown specimen. Its habitat is mainly forests, but it has also been recorded in dense savannah biotopes. Though widespread throughout Africa it is not particularly numerous. It has a slow lazy flight pattern.

Elsewhere larvae feed on *Secamone* species, *Tylophora* species and *Heliotropium indicum*, all of which occur in Seychelles. However, this butterfly is not considered a Seychelles resident species. This taxon is under no threat.

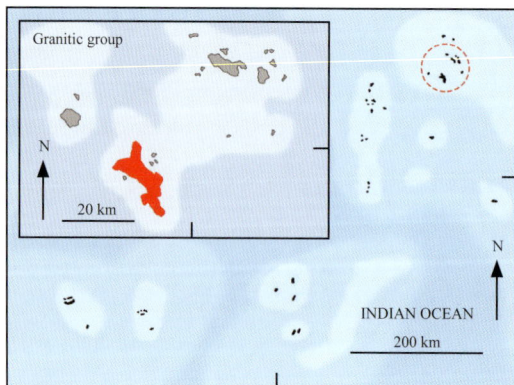

Granitic group

N

20 km

N

INDIAN OCEAN

200 km

Above top: *A. niavius dominicanus*
Above bottom: *A. niavius dominicanus* mud-puddling
Left: Seychelles distribution of *A. niavius dominicanus*

Euploea rogeri (Geyer, 1837)

Seychelles endemic
Extinct

Plate 6
Crow

Description

♀ UPF black-brown ground colour with cream-white patches in apex. UPH black-brown ground colour with numerous large cream-white patches in post-median area. UNF and UNH similar colour pattern to wing dorsal surfaces. Thorax with white spots. ♂ is unknown.

Hübner 1837

Distribution

Endemic to Seychelles. Assumed to have been collected on Mahé.

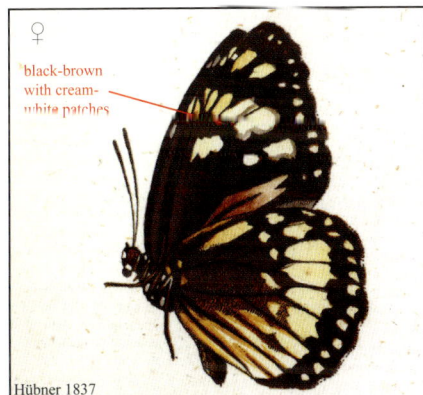

black-brown with cream-white patches

Hübner 1837

Biology and conservation

Described from a specimen of unknown origin dating from c. 1820, suggested by de Joannis (1894) to be from Seychelles. The type specimen is lost and it is now only known from two illustrations by Geyer in Hübner (1837). It may be synonymous with *Euploea mitra* (Legrand 1965). Lawrence (2009b) suggested it could be an extinct subspecies of the Mascarene *Euploea euphon*, based on a similarity in wing colour pattern. This butterfly, if it existed at all, is considered 'Extinct'.

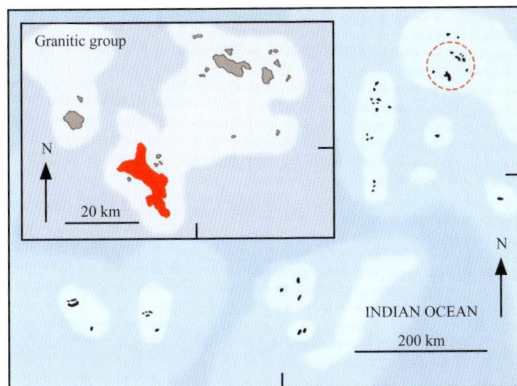

Above: *E. rogeri*
Left: Assumed distribution of *E. rogeri* in Seychelles

Euploea mitra Moore, 1858

Seychelles endemic
Endangered

Plate 6

Seychelles Crow

Description

Sexes similar. LFW: ♂ 70–78 mm. ♀ 66–74 mm. UPF black with large white patches in the post-median and apical areas. UPH black ground colour with numerous faint white spots along margins. UNF and UNH similar in colour pattern to dorsal wing surfaces, but with more distinct white spots on UNH. Thorax has numerous white spots. ♂ has distinct sex-brand and a curved FW trailing edge. ♀ lacks the sex-brand and has a straight FW trailing edge.

sex-brand

♂

curved wing margin

straight wing margin

♀

Distribution

Endemic to Seychelles where it has been recorded on the islands of Mahé, St. Anne, Silhouette and La Digue.

Biology and conservation

The Seychelles Crow flies slowly, with both sexes attracted to flowers. It is occasionally seen at sea-level, but is usually observed at higher altitudes. Early stages are unknown, but the larval food plant is thought to be *Tylophora* species or *Heliotropium indicum* (Gerlach & Matyot 2006).

It has declined dramatically over the last 150 years and is rarely seen now. Most sightings are during March and April. It is listed as 'Endangered', based on a decline in numbers and small range.

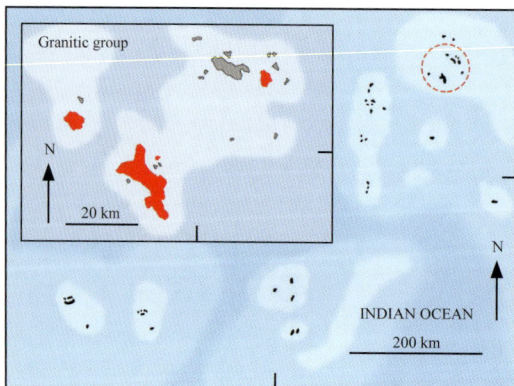

Granitic group

N

20 km

INDIAN OCEAN

200 km

N

Above: *E. mitra*

Left: Seychelles distribution of *E. mitra*

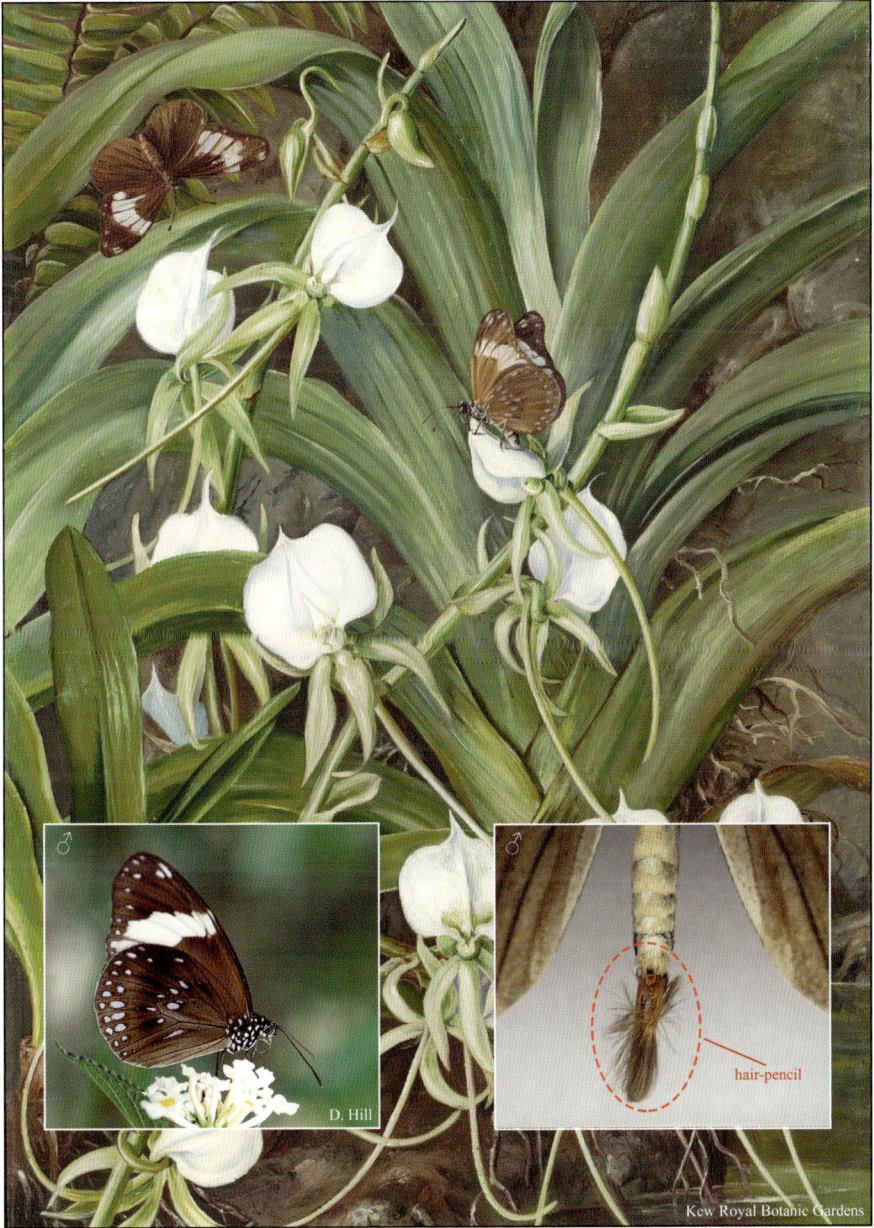

Above: Painting 489 (c. 1883) by Marianne North, depicting *E. mitra* on a Seychelles orchid
Inset left: *E. mitra;* **Inset right**: Hair-pencil on male *E. mitra* abdomen

Acraea neobule Doubleday, 1847

Afrotropical region
Not threatened

Plate 7

Dancing Acraea

Description

Sexes similar. LFW: ♂ 48–55 mm. ♀ 50–56 mm. UPF orange with transparent post-median area. Numerous black spots in discal cell. UPH ground colour orange with black margin and numerous black spots in basal and post-median areas. The UNF and UNH have similar markings to the dorsal surfaces except pink-red in colour. ♀ slightly paler in overall colour. Additional image: page 94.

large transparent area

orange ground colour

Distribution

The Afrotropical region including Comoros and Seychelles, where it occurs on Aldabra, Assumption, Cosmoledo and Astove.

Biology and conservation

Change in the plant community composition on Aldabra since 1976 has resulted in its extirpation on that island (Gerlach 2012).

The larvae have been recorded on *Turnera angustifolia* (Legrand 1965).

This butterfly was originally classified as *Acraea terpsicore legrandi* which was considered endemic to Seychelles (Carcasson 1964). However, *Acraea terpsicore* occurs in Asia, with the African populations classified as *Acraea neobule* (Henning 1986; Pierre & Bernaud 2013).

The subspecies *legrandi* was described as being similar to the African individuals, but differed by being pinker in ground colour and with smaller black spots. However, *neobule* is quite variable in colour and pattern markings (D. Bernaud *pers. comm.*), with the Seychelles specimens indistinguishable from those from South Africa, Comoros and Socotra. Not threatened.

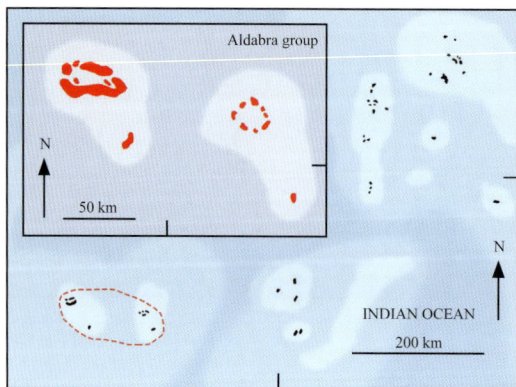

Aldabra group

N

50 km

INDIAN OCEAN

N

200 km

Above: *A. neobule*

Left: Seychelles distribution of *A. neobule*

Acraea ranavalona Boisduval, 1833

Malagasy subregion
Not threatened

Plate 7
Ranavalona's Acraea

Description

Sexes dimorphic. ♂ LFW: 38–46 mm. UPF mostly transparent, except for red dusting in basal and median areas. UPH ground colour red with numerous black spots in the post-median and terminal areas. UNF and UNH similar in colour pattern to the dorsal surfaces except the red colour is paler. ♀ LFW: 51–53 mm. ♀ polymorphic. One form is similar to ♂ but with the red replaced by a pale-white-red ground colour. The other ♀ form is red and resembles the ♂.

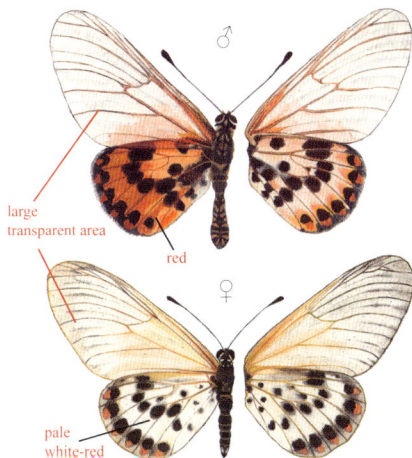

large transparent area

red

pale white-red

Above: *A. ranavalona*

Distribution

Madagascar, Comoros and Seychelles, where it occurs on the islands of Aldabra and Astove.

Biology and conservation

This butterfly was reported to be abundant on Aldabra in the 1970s, but is now rarely seen (Gerlach 2012). Larvae have been recorded on *Tylophora coriacea* and *Passiflora suberosa*. Adults seen between December and March. It is not threatened at present.

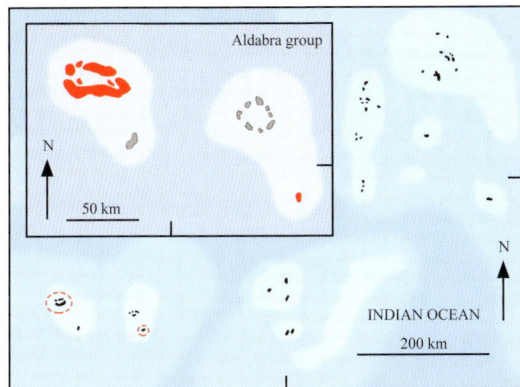

Above: *A. ranavalona* mating pair
Left: Seychelles distribution of *A. ranavalona*

56

Phalanta phalantha aethiopica (Rothschild & Jordan, 1903)

Afrotropical region
Not threatened

Plate 7

African Leopard

Description

Sexes similar. LFW: ♂ 40–45 mm. ♀ 43–48 mm. UPF ground colour orange with black patches in terminal, apical and discal areas. UPH orange ground colour with black marginal border. UNF and UNH ground colour variable from purple-brown (rare) to orange-brown (common). Similar spot and marking pattern to that on wing dorsal surfaces, but less distinct. Similar to *Phalanta philiberti* but FW more rounded. Additional image: Plate 7.

Distribution

The Afrotropical region including the Indian Ocean islands. In Seychelles it occurs on Aldabra and Astove.

Biology and conservation

The African Leopard is an inhabitant of savannah and woodland biotopes. Its flight is often fast and erratic. Larsen (2005) reports this species as migratory.

The larvae are polyphagous and in Seychelles they have been recorded on *Flacourtia indica* (Matyot 2002). This butterfly is widespread across Africa, common and under no threat.

orange ground colour

♂

South Africa   S. Woodhall

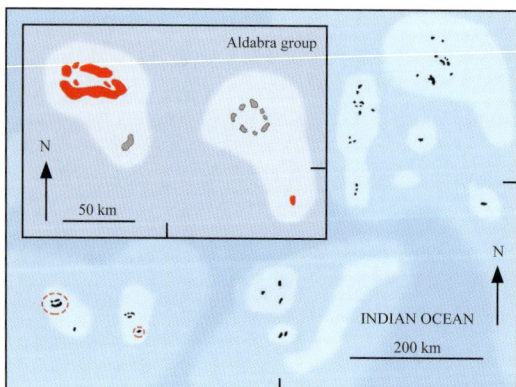

Aldabra group

N

50 km

N

INDIAN OCEAN

200 km

Above top: *P. phalantha aethiopica*
Above bottom: *P. phalantha aethiopica* feeding
Left: Seychelles distribution of *P. phalantha aethiopica*

Phalanta philiberti (de Joannis, 1893)

Seychelles endemic
Extinct

Plate 7

Seychelles Leopard

Description

Sexes similar. LFW: 48–56 mm. UPF ground colour orange-brown with well-defined black markings. Basal area rich red-brown. UNH similar in colour and markings to UPF, but with the rich red-brown scales extending across the basal and median areas. UNF and UNH ground colour metallic-ochreous-green with similar spots and patterns as found on the dorsal surfaces, but less distinct. Similar to *Phalanta phalantha* but larger, with well-defined markings and pointed FWs.

Distribution

Endemic to the Seychelles granitic islands where it was recorded on the islands of Mahé, Praslin and Silhouette.

Biology and conservation

The only information on this species was

brown-orange ground colour

given by Fletcher (1910 page 292): *"It is found at a height of about 800–1000 feet and over, but is not seen lower down. It is fond of beds of streams, where it flies over the trees in company with* Euploea mitra, *having an elegant sailing flight. In Praslin examples, as Joannis has remarked, the metallic patches of the underside are usually more yellowish than green, but this is paralleled in many of the Mahé specimens and so cannot be taken as a distinctive race character"*. Fryer (1912) reported it as abundant in 1908–1909. The early stages of this species were never recorded.

This butterfly was last collected in 1953 with more individuals seen between 1956 and 1960 (Legrand 1965). Unfortunately it is now considered 'Extinct'.

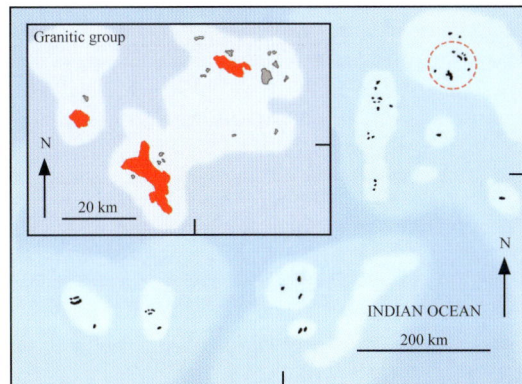

Granitic group

N

20 km

N

INDIAN OCEAN

200 km

Above: *P. philiberti*
Left: Seychelles distribution of *P. philiberti*

Melanitis leda helena (Westwood, 1851)

Afrotropical region

Plates 8 and 9

Not threatened

Twilight Brown

overall
brown
colour

intermediate

♂

♀

distinct
ocelli

w.s.f.

♂

d.s.f. aberration

♂

poorly
defined
ocelli

d.s.f.

Description

Sexes similar. LFW: ♂ 58–65 mm. ♀ 63–72 mm. Overall brown ground colour on all wing surfaces. Two seasonal forms occur with intermediates. The w.s.f. has blunt wingtips and large prominent ocelli. The ventral surface is uniform with an irroration of brown and white. The d.s.f. has a variegated wing margin with pointed wingtips and reduced ocelli. Wing ventral surfaces are strongly cryptic and variable. The ♂ is generally darker than the ♀.

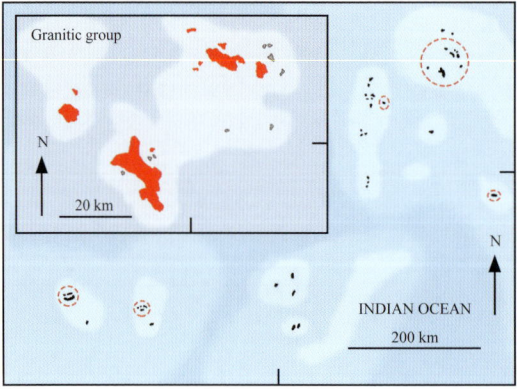

Granitic group

N

20 km

N

INDIAN OCEAN

200 km

Above: *M. leda helena*
Left: Seychelles distribution of *M. leda helena*

Distribution

Afrotropical region including the Indian Ocean islands. In Seychelles it has been recorded on Mahé, St. Anne, Praslin, Silhouette, North Island, La Digue, Aride, Cousin, Cousine, Curieuse, Coëtivy, Desroches, Aldabra and Cosmoledo.

Biology and conservation

The Twilight Brown is abundant in forest and woodland biotopes. On the granitic islands it is often attracted to fallen *Artocarpus heterophyllus* fruit. This species is crepuscular and regularly attracted to lights. They do not often visit flowers.

In Seychelles, the Twilight Brown flies during most months of the year. However, it is more often seen during the wet NW Monsoon period. Larvae are polyphagous on the Poaceae and Palmae of which there are numerous species in Seychelles.

Larsen (2005) does not consider the African population subspecifically distinct from the Oriental nominate population. This butterfly is very widespread, common and under no threat.

P. Mazzei

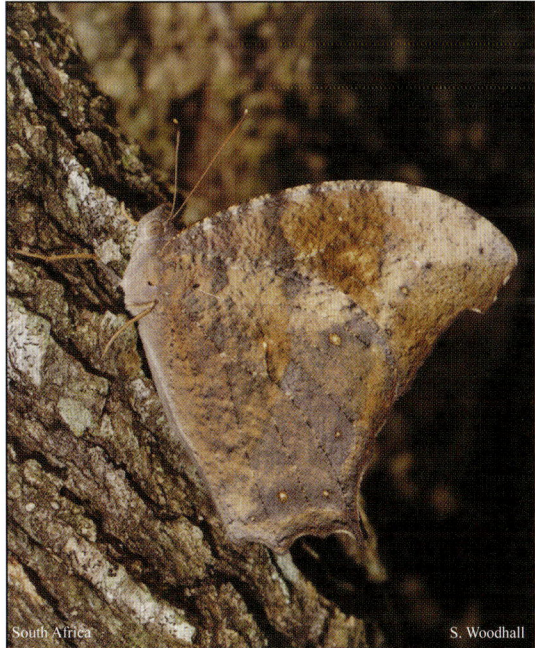
South Africa
S. Woodhall

Top: *M. leda helena* w.s.f. perching
Bottom: *M. leda helena* d.s.f. camouflaged on a tree trunk

Hypolimnas misippus (Linnaeus, 1764)

Cosmopolitan

Not threatened

Plate 10

Diadem Eggfly

large white patch

white bands

intermediate

no black spots

f. *inaria*

f. *misippus*

Description

Sexes dimorphic. ♂ LFW: 60–65 mm. UPF and UPH velvet-black ground colour with large metallic-blue-bordered white patches in the centre of each wing. The metallic-blue border is only visible at oblique angles.

UNF red-brown basal area, large white transverse white band in median area, light-brown post-median area with a small white spot in apical area. Terminal area with black and white margin. UNH with well developed white band in

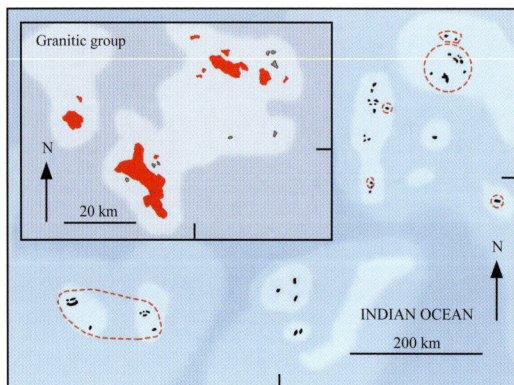

Above: *H. misippus*

Left: Seychelles distribution of *H. misippus*

the median area. Basal and post-median areas light-brown. Tornal area red-brown. Terminal area with black and white margin. ♀ LFW: 70–80 mm. ♀ polymorphic and closely resembles *Danaus chrysippus* and *D. dorippus* (pages 49–50) which she mimics. The simplest way to differentiate ♀ *H. misippus* from the *Danaus* species is by the absence of the three black spots bordering the HW discal cell in *H. misippus*. Additional images: page 21 and 124.

Distribution

Cosmopolitan in distribution. In Seychelles it has been recorded on the islands of Mahé, St. Anne, North Island, Praslin, Aride, La Digue, Silhouette, Marianne, Cousin, Cousine, Curieuse, Denis, Bird, Coëtivy, Desroches, Alphonse, Aldabra, Assumption, Astove and Cosmoledo.

Biology and conservation

The Diadem Eggfly is the most conspicuous butterfly in the granitic island group. Butterflies appear suddenly, with numbers increasing rapidly (Bourquin et al. 2000; Lawrence 2004b). They

South Africa — S. Woodhall

South Africa — S. Woodhall

are on the wing between December and March (Lawrence 2005). Peaks in butterfly abundance are mirrored by similar peaks in rainfall, with rainfall peaks occurring a few days before those of the butterfly (Lawrence 2000b).

Flight is more active than in the *Danaus* species. This butterfly flies in open areas and along forest margins. Larval food plants include *Asystasia* species and *Portulaca oleracea*. This species is not threatened.

Top: Male *H. misippus* feeding; **Bottom**: Female *H. misippus* feeding

Hypolimnas bolina jacintha (Drury, 1773)

Oriental region
Not threatened

Plate 11
Great Eggfly

Description

Sexes dimorphic. ♂ LFW: 65–70 mm. Similar to *Hypolimnas misippus* ♂ but larger, with smaller blue-bordered pale-blue patches on wing dorsal surfaces. UNF and UNH brown-black ground colour with faint brown-white bands in terminal and post-median areas. ♀ LFW: 80–86 mm. UPF and UPH brown-black ground colour with cream-white margins, which are broader and more extensive on the HW than on the FW. UNF and UNH have similar markings and patterns as found on the wing dorsal surfaces.

♂

pale-blue patch

brown-black colour

♀

cream-white margin

brown-black colour

Distribution

Widespread across the Oriental region. In the Afrotropical region this butterfly has been recorded on Madagascar (Paulian 1956), Socotra (Ogilvie-Grant 1903), Kenya, Ethiopia (Larsen 1996), Mauritius (Williams 2007) and Seychelles. There are three records for Seychelles. These are from Mahé (Lawrence 2011), Alphonse (Betts 2009) and an unconfirmed sighting on Marianne (Hill et al. 2002).

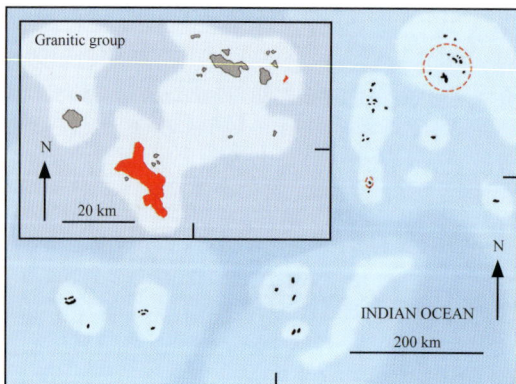

Granitic group

N

20 km

N

INDIAN OCEAN

200 km

Biology and conservation

Elsewhere, larvae are polyphagous on Acanthaceae and Malvaceae, which occur in Seychelles. This butterfly is widespread outside Seychelles and not threatened.

Above: *H. bolina jacintha*
Left: Seychelles distribution of *H. bolina jacintha*

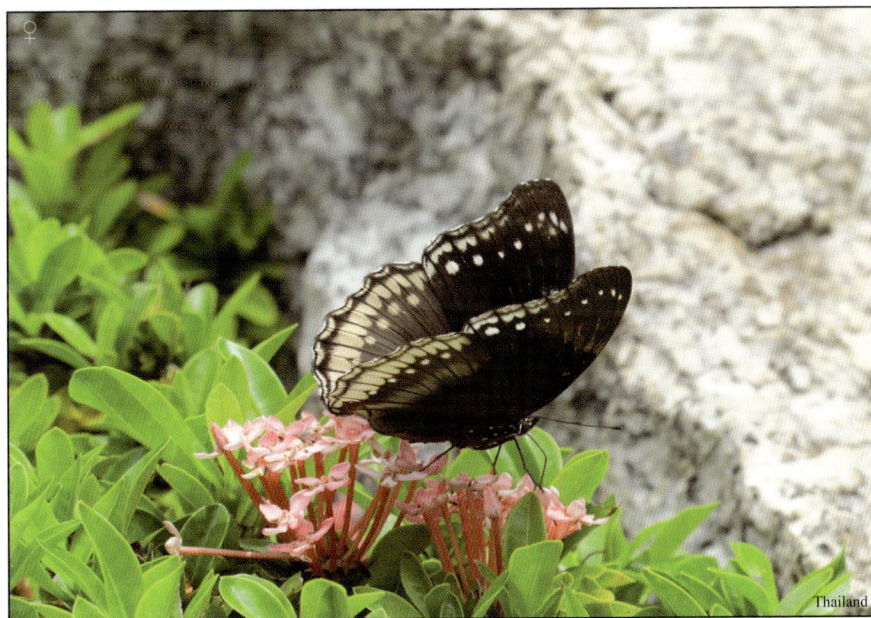

Top left: Male *H. bolina jacintha* perching

Top right: Female *H. bolina jacintha* feeding

Bottom: Female *H. bolina jacintha* feeding

Junonia rhadama (Boisduval, 1833)

Malagasy subregion
Not threatened

Plate 12

Royal Blue Pansy

distinct ocelli

poorly defined ocelli

w.s.f.

d.s.f.

one ocellus

♂

♀

two ocelli

Description

Sexes dimorphic. ♂ LFW: 47–49 mm. UPF bright-blue, with brown and white terminal margins. Small white spots in apcx. UPH bright-blue with single ocellus between CuA_2 and 1A+2A. Wing with white terminal margin. UNF and UNH with irroration of brown, white and black. ♀ LFW: 48–51 mm. Similar to the male, but with more distinct markings (especially the white spots on the FW) on the wing dorsal surfaces. ♀ has two ocelli on the UPH between CuA_2 and 1A+2A, and dissecting Rs. Ventral surfaces same as in the ♂.

There are two seasonal forms. The w.s.f. has prominent ocelli and more pointed wingtips, whereas the d.s.f. has reduced ocelli, blunter wingtips and more cryptic ventral surface colours and patterns.

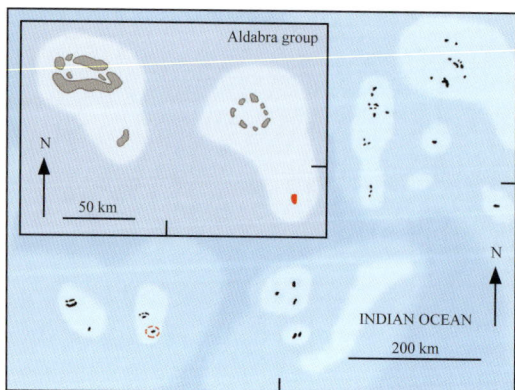

Aldabra group

N

50 km

N

INDIAN OCEAN

200 km

Above: *J. rhadama*

Left: Seychelles distribution of *J. rhadama*

65

Distribution

Mauritius, Réunion, Madagascar, Comoros and Seychelles, where it occurs on Astove.

Biology and conservation

This butterfly is not widely collected in Seychelles. Known only from six specimens collected in the early 1900s (Fryer 1912) and from sightings in the 1960s (Beamish 1970).

It is an active and wary insect that is difficult to approach. Males are territorial and often settle on the ground or a rock, where their cryptic ventral surfaces are well camouflaged. This butterfly is on the wing during the wet NW Monsoon period.

The larvae feed on *Barleria lupulina* on Mauritius (Williams 2007) and *B. prionitis* on Réunion (Martiré & Rochat 2008). This species is widespread in the Malagasy subregion and not threatened.

Top right and left: Male *J. rhadama* perching
Below: Female *J. rhadama* perching

Junonia oenone epiclelia (Boisduval, 1833)

Malagasy subregion

Plate 13

Not threatened

Dark Blue Pansy

large ocelli

no blue patch

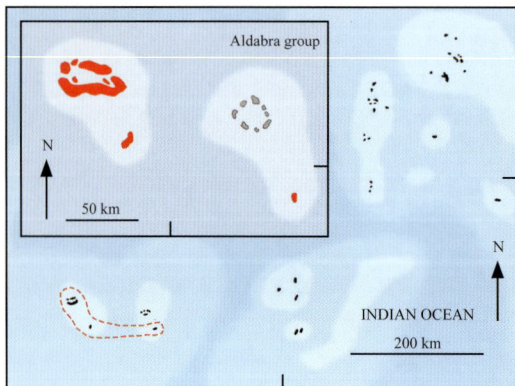

small ocelli

distinct blue patch

large ocelli

indistinct blue patch

Description

Sexes dimorphic. ♂ LFW: 40–48 mm. UPF ground colour black with a distinct transverse series of white blocks in median and post-median areas. Two small ocelli in terminal area along with several white spots. Two orange-brown lines in discal cell. UPH black ground colour with a large metallic-blue patch in centre of the wing. Two small ocelli in terminal area. Wing margin black and white. UNF variable in colour, but usually red-brown ground colour

with similar marking as on UPF. Basal area red-brown with two black-bordered white stripes. Apex mottled brown. Terminal and post-median areas black with white patches. UNH mottled brown with a red-brown transverse line through the post-median area and with two small ocelli. ♀ LFW: 46–52 mm. ♀ polymorphic with two forms, and similar to ♂. In one form the black ground colour on the wing ventral surfaces is replaced by brown, the

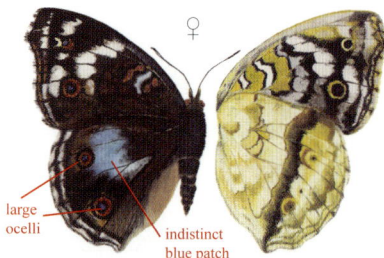

Aldabra group

N

50 km

N

INDIAN OCEAN

200 km

Above: *J. oenone epiclelia*

Left: Seychelles distribution of *J. oenone epiclelia*

ocelli are larger and the blue patch on UPH absent. In the other form the ground colour is brown-black, the ocelli larger and with an indistinct large metallic-blue patch on UPH.

Distribution

Madagascar and Seychelles, where it occurs on Aldabra, Assumption and Astove.

Biology and conservation

The Dark Blue Pansy is common in the Aldabra island group. Adults are on the wing during the NW Monsoon, but it is most abundant from February to April.

On Madagascar the larvae feed on *Hypoestes verticillaris* (Paulian 1956). *Hypoestes aldabrensis* is endemic to the Aldabra group and is the only *Hypoestes* species in Seychelles. This plant may be the food plant for this taxon in Seychelles. This butterfly is widespread and abundant on Madagascar. Not threatened.

Above: *J. oenone epiclelia* perching

Junonia orithya madagascariensis (Guenée, 1863)

Afrotropical region

Not threatened

Plate 13

Eyed Pansy

Description

Sexes similar. LFW: ♂ 35–42 mm. ♀ 40–48 mm. UPF basal area black. Tornal area metallic-blue. Two large ocelli in terminal area. UPH metallic-blue with black basal area. Two large ocelli in terminal area. UNF and UNH variable but usually light-brown ground colour with ocelli in terminal areas. UNF has prominent yellow-orange, black and white stripe pattern in discal cell. ♀ similar to ♂ but with duller metallic-blue. Additional image: page 124.

metallic-blue

Distribution

Afrotropical region including Madagascar, Comoros and Seychelles, where it occurs on Aldabra, Assumption and Cosmoledo.

dull metallic-blue

Biology and conservation

The Eyed Pansy is a savannah butterfly with migratory tendencies (Larsen 2005). Butterflies are on the wing during the NW Monsoon and usually most abundant from February to April. There are no current Seychelles records, with the most recent being from Aldabra in 1954 (Gerlach & Matyot 2006).

Elsewhere, the larvae feed on various species of Acanthaceae and Verbenaceae. It is widespread across Africa and not threatened.

Aldabra group

N

50 km

N

INDIAN OCEAN

200 km

Above: *J. orithya madagascariensis*
Left: Seychelles distribution of *J. orithya madagascariensis*

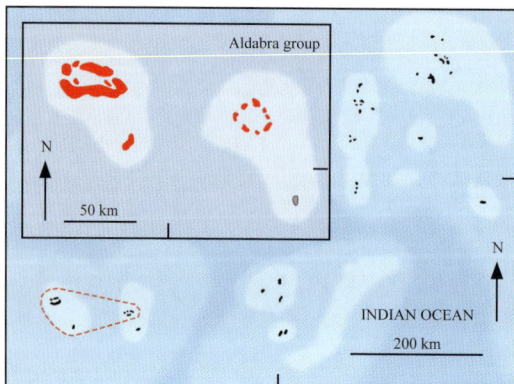

Top: Male *Junonia orithya madagascariensis* perching

Bottom: Female *Junonia hierta cebrene* feeding (see *Species Accounts* page 71)

Junonia hierta cebrene (Trimen, 1870)

Afrotropical Region

Plate 13

Not threatened

Yellow Pansy

Description

Sexes similar. LFW: ♂ 40–45 mm. ♀ 40–50 mm. UPF black with large yellow patch in centre of wing. Small ocellus between CuA_2 and 1A+2A. UPH black with large yellow patch in tornal area. Large blue spot towards leading edge of wing. UNF brown. Centre of wing yellow with small ocellus between CuA_2 and 1A+2A. Discal cell with yellow-brown, black and white stripes. UNH brown ground colour. ♀ similar to ♂, but with larger ocelli and a black dot in yellow tornal area on UPH. Additional image: page 70.

yellow patches

♂

♀

black dot

Distribution

The African continent. In Seychelles it was recorded from Île du Sud-est. Île du Sud-est was a separate islet neighbouring Mahé until 1971. It is now part of Mahé mainland due to land reclamation for the international airport.

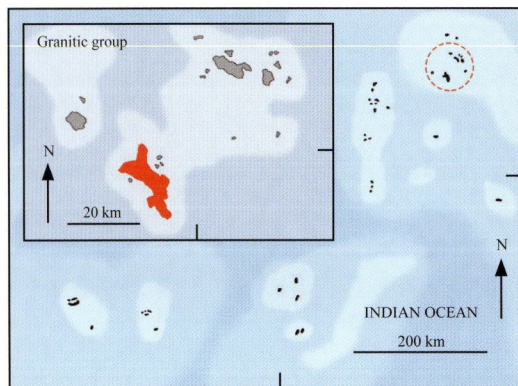

Granitic group

N

20 km

INDIAN OCEAN

N

200 km

Biology and conservation

So far the Yellow Pansy is known only from a single capture by M. Mason in 1953 (Legrand 1965). It may represent a temporary population or a wind-blown vagrant specimen.

Elsewhere, the larvae feed on various species of Acanthaceae. This butterfly is not considered a Seychelles resident species. Under no threat.

Above: *J. hierta cebrene*
Left: Seychelles distribution of *J. hierta cebrene*

71

Vanessa cardui (Linnaeus, 1758)

Cosmopolitan

Not threatened

Plate 12

Painted Lady

Description

Sexes similar. LFW: ♂ 40–45 mm. ♀ 45–50 mm. UPF large black apical area with numerous white spots and patches. Basal and median areas red-orange with black patches. Terminal area black. UPH red-orange. Numerous black spots in post-median and terminal areas. UNF similar colour and pattern as UPF, but with red-orange colour more red. UNH mottled white-brown with numerous ocelli between the veins in the terminal area. Veins white. Additional image: page 18.

large black apical area

red-orange colour

South Africa — S. Woodhall

Distribution

Cosmopolitan, including the Indian Ocean islands. In Seychelles it occurs on Mahé, St. Anne, Praslin, Silhouette, La Digue, Aride, Cousine, Bird Island, Coëtivy, Desroches, Aldabra and Assumption.

Biology and conservation

Although widespread in Seychelles it is not common. The Painted Lady is a well-known migrant. Gerlach & Matyot (2006) suggest that this species appears as a non-breeding migrant. Recorded during December and January. Larvae feed on numerous plant families. Not threatened.

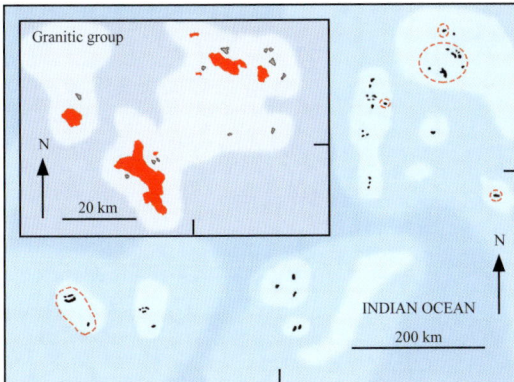

Granitic group

N

20 km

INDIAN OCEAN

N

200 km

Above top: *V. cardui*
Above bottom: *V. cardui* feeding
Left: Seychelles distribution of *V. cardui*

Lycaenidae

The Lycaenidae, commonly known as the blues and hairstreaks, form a huge and diverse cosmopolitan family. Worldwide, there are an estimated 6000 species in seven subfamilies (Larsen 2005), with approximately 1700 Afrotropical species. Two subfamilies, the Theclinae and the Polyommatinae, occur in Seychelles.

Although, as a family, they are varied in colour pattern, they are all generally small butterflies. Many species are brightly coloured. The forelegs, especially in the males, are reduced as in the Nymphalidae. All Seychelles species are blue, at least in the males. Also, many lycaenid species are myrmecophilous. These relationships vary from casual to obligatory.

Six species occur in Seychelles. Two other species have been erroneously associated with Seychelles and warrant further discussion:

1) *Euchrysops malathana* was listed as occurring on Aldabra (Holland 1896), but was identified from worn specimens and mostly likely confused with the similar *Euchrysops osiris*.

2) A single *Zizeeria maha* specimen was apparently collected on Mahé by H.P. Thomasset (Fletcher 1910). This is an Oriental species and was probably confused with the similar looking *Zizeeria knysna*.

The Seychelles Lycaenidae subfamilies

Theclinae: About 2800 species with a worldwide distribution. There are roughly 520 species in the Afrotropical region. Larvae feed on many plant families.

Polyommatinae: Worldwide distribution with roughly 1400 species. There are about 450 Afrotropical species. Larvae feed on a large variety of plant families.

Simple key to the Seychelles Lycaenidae species

1. Hindwings with tails 2
   Hindwings without tails 5

2. Hindwings with two tails .... *Hypolycaena philippus ramonza*
   Hindwings with one tail 3

3. Ventral wing surfaces ground colour white .... *Euchrysops osiris*
   Ventral wing surfaces ground colour not white 4

4. Ventral wing surfaces fawn ground colour .... *Lampides boeticus*
   Ventral wing surfaces dark-grey and white .... *Leptotes pirithous*

5. Ventral surface wing terminal areas with black dots .... *Zizeeria knysna*
   Ventral surface wing terminal areas with black dashes .... *Zizula hylax*

Hypolycaena philippus ramonza (Saalmüller, 1878)

Malagasy subregion

Plate 14

Not threatened

Purple Brown Hairstreak

Description

Sexes dimorphic. ♂ LFW: 25–28 mm. The only lycaenid in Seychelles with two tails. UPF and UPH purple ground colour. UPH with two large ocelli in tornal area. UNF and UNH ground colour light-grey-white with three orange-brown transverse bands. UNH with three broken transverse bands and two ocelli in tornal area. ♀ LFW: 26–29 mm. UPF grey-brown. UPH grey-brown ground colour, with white patches in post-median area and two ocelli in tornal area. The ventral surfaces are the same as in the ♂.

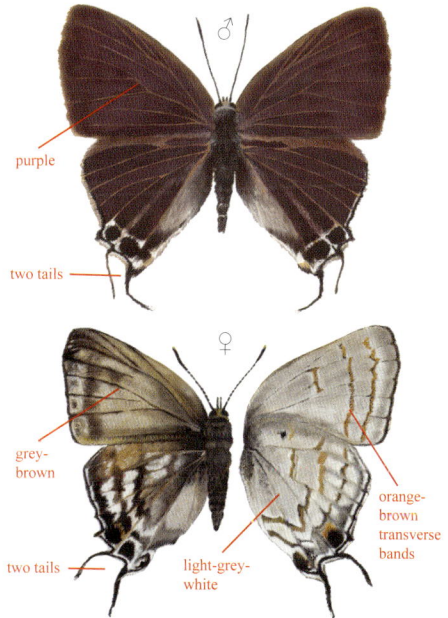

Distribution

Madagascar, Comoros and Seychelles, where it occurs on Aldabra and Cosmoledo.

Biology and conservation

The Purple Brown Hairstreak is a fast flying species with highly territorial males. Both sexes are attracted to flowers. On Madagascar this butterfly is an inhabitant of forested areas, with the capacity to colonise degraded forests (Lees et al. 2003).

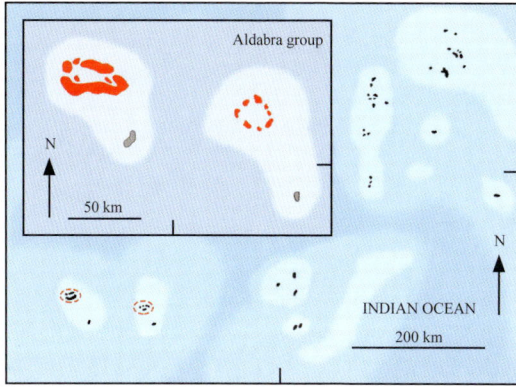

The early stages of this taxon are unknown. This butterfly is widespread on Madagascar and under no immediate threat.

Above: *H. philippus ramonza*
Left: Seychelles distribution of *H. philippus ramonza*

Lampides boeticus (Linnaeus, 1767)

Cosmopolitan

Not threatened

Plate 14

Long Tailed Blue

Description

Sexes dimorphic. ♂ LFW: 24–32 mm. UPF and UPH purple-blue with black FW margin. Leading edge of UPH with broad light-brown margin. UPH with single tail and two ocelli in tornal area. UNF and UNH fawn-brown ground colour, with brown and pale-white transverse stripes. Two distinct ocelli in tornal area of UNH. ♀ LFW: 24–34 mm. UPF and UPH with blue basal and median areas. Post-median and terminal areas of both wings brown. HW tailed with two distinct ocelli in tornal area. Ventral surfaces are the same as in the ♂.

purple-blue

one tail

brown

♂

blue

one tail

fawn-brown

♀

brown and pale-white transverse bands

Distribution

Cosmopolitan, including the Indian Ocean islands. In Seychelles it occurs on Mahé, St. Anne, Silhouette, Praslin, La Digue, Cousine, Cousin, Aride, Curieuse and Cosmoledo.

Granitic group

N

20 km

N

INDIAN OCEAN

200 km

Biology and conservation

The Long Tailed Blue is a commonly encountered butterfly in Seychelles. It is found in open habitats including forest edges. It is a strong, fast flyer and migratory. Adults are on the wing throughout the year.

Larvae are polyphagous, and have been recorded feeding on *Crotalaria* species in Seychelles. This butterfly is widespread and under no threat.

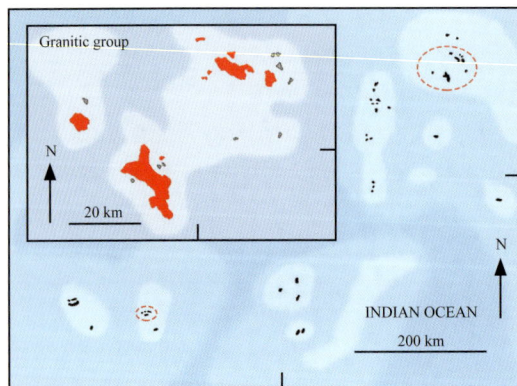

Above: *L. boeticus*

Left: Seychelles distribution of *L. boeticus*

South Africa — S. Woodhall

South Africa — S. Woodhall

Top: Male *L. boeticus* perching
Bottom: Female *L. boeticus* perching

Leptotes pirithous (Linnaeus, 1767)

Cosmopolitan
Not threatened

Plate 14

Common Blue

Description

Sexes dimorphic. ♂ LFW: 21–29 mm. UPF and UPH purple-blue with a single tail on the HW. There are no ocelli in the UPF tornal area. UNF and UNH with white and dark-grey transverse stripes. UNH with two ocelli in tornal area. ♀ LFW: 24–30 mm. UPF and UPH basal and median areas blue, with mottled grey and white median and terminal areas. Wing margins black. Single tail on HW, with two ocelli in tornal area. Ventral surfaces are the same as in the ♂.

♂

no ocelli

one tail

♀

white and dark-grey transverse stripes

one tail

Distribution

Cosmopolitan, including the Indian Ocean islands. In Seychelles it has been recorded on Mahé, Cerf Island, St. Anne, Silhouette, Praslin, La Digue, Cousin, Aride, Curieuse, Bird Island, Rémire, St. Pierre, Aldabra, Assumption, Astove and Cosmoledo.

Biology and conservation

A widespread and common Seychelles butterfly that is found in open biotopes and disturbed areas in forests. This species is migratory. Adults are on the wing throughout the year. Both sexes are attracted to flowers.

The larvae are polyphagous on numerous plant families including the Papilionaceae, the Labiatae, the Oleaceae and the Plumbaginaceae, all of which occur in Seychelles. The Common Blue is widespread, abundant and not threatened.

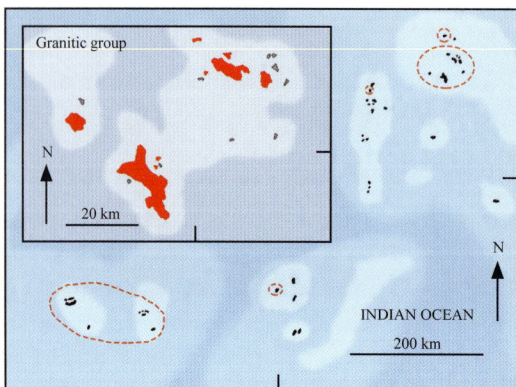

Granitic group

N

20 km

INDIAN OCEAN

N

200 km

Above: *L. pirithous*
Left: Seychelles distribution of *L. pirithous*

77

Top: Male *L. pirithous* perching
Bottom: Female *L. pirithous* perching

Zizeeria knysna (Trimen, 1862)

Afrotropical region

Not threatened

Plate 14

Sooty Blue

Description

Sexes dimorphic. ♂ LFW: 18–23 mm. UPF and UPH blue ground colour with broad dark-brown to black margins. UNF and UNH fawn colour with a series of black dots in the median and terminal areas of both wings. ♀ LFW: 17–25 mm. UPF and UPH brown to dark-brown ground colour, with variable dusting of blue scales in basal areas. This blue dusting sometimes extends into the median area. Ventral surfaces are the same as in the ♂. Additional image: page 94.

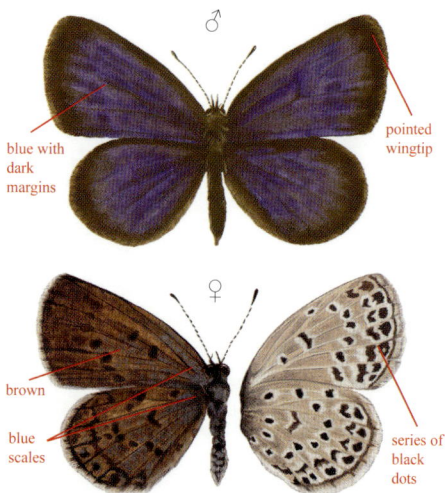

♂

blue with dark margins

pointed wingtip

♀

brown

blue scales

series of black dots

Distribution

Cosmopolitan, including the Indian Ocean islands. In Seychelles it occurs on Mahé, Cerf Island, St. Anne, Silhouette, Praslin, La Digue, Cousin, Cousine, Aride, Curieuse, Bird Island, Denis Island, Desroches, Rémire, Farquhar and Aldabra.

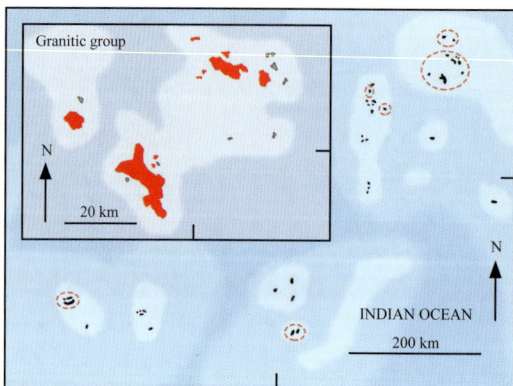

Granitic group

N

20 km

N

INDIAN OCEAN

200 km

Biology and conservation

The Sooty Blue is the most commonly encountered butterfly in Seychelles. It flies in open areas and does not occur in forest biotopes (Lawrence 2009c). It flies close to the ground, seldom venturing more than half a metre above the ground. Adults are on the wing throughout the year.

The larvae are polyphagous, and have been observed feeding on *Amaranthus dubius*. This species is under no threat.

Above: *Z. knysna*
Left: Seychelles distribution of *Z. knysna*

Top left: *Z. knysna* perching
Top right: Female *Z. knysna* perching
Bottom: Male *Z. knysna* feeding

Zizula hylax (Fabricius, 1775)

Cosmopolitan
Not threatened

Plate 14
Gaika Blue

Description

Sexes dimorphic. ♂ LFW: 17–21 mm. UPF and UPH pale-blue with light-brown margins. UNF and UNH fawn coloured with a series of faded brown and white dashes in the terminal areas of both wings. Basal areas of both wings have a dusting of black scales. ♀ LFW: 18–25 mm. UPF and UPH dark-brown. Ventral surfaces are the same as in the ♂. Additional image: page 124.

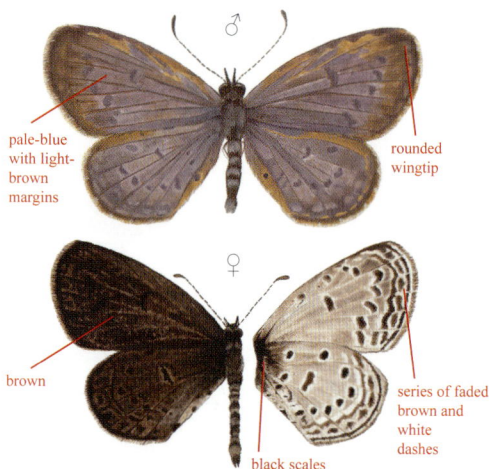

pale-blue with light-brown margins

rounded wingtip

brown

black scales

series of faded brown and white dashes

Distribution

Cosmopolitan, including the Indian Ocean islands. In Seychelles it occurs on Mahé, Silhouette, Praslin, La Digue, Aride, Alphonse, Providence and Aldabra.

Biology and conservation

The Gaika Blue is another widespread and common butterfly in Seychelles. Its flight is weak, usually among the undergrowth rather than just above the ground as in *Zizeeria knysna*. It only occurs in open biotopes and is not found in forested areas. This butterfly is seldom observed at higher elevations. It is often seen in gardens and disturbed areas. Both sexes are attracted to flowers, in particular *Lantana camara*. The males are often seen mud-puddling in large numbers.

Larvae are polyphagous on numerous plant families, including the Poaceae and Mimosaceae. This species not threatened.

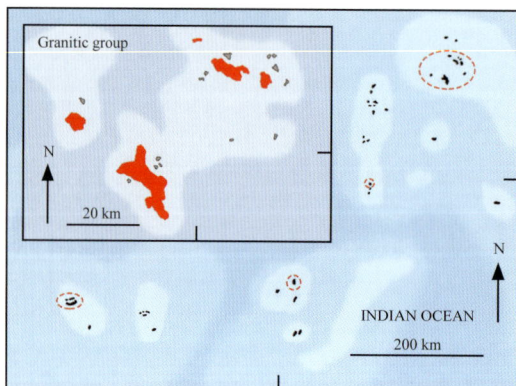

Granitic group

N

20 km

N

INDIAN OCEAN

200 km

Above: *Z. hylax*

Left: Seychelles distribution of *Z. hylax*

Top: Male *Z. hylax* perching
Bottom: Female *Z. hylax* perching

Euchrysops osiris (Hopffer, 1855)

Afrotropical region

Not threatened

Plate 14

Osiris Smokey Blue

Description

Sexes dimorphic. ♂ LFW: 22–29 mm. UPF and UPH pale-brown-blue with black margins and a single tail on the HW. UPH with two small ocelli in the tornal area. Basal areas with dark-blue scales. UNF and UNH white with grey transverse stripes in post-median and terminal areas. Basal areas with black scales. UNH with three black spots. ♀ LFW: 25–30 mm. UPF and UPH mottled grey-brown with bright-blue centre. UPH has white zigzag line in post-median area with two ocelli in tornal area. Ventral surfaces are the same as in the ♂.

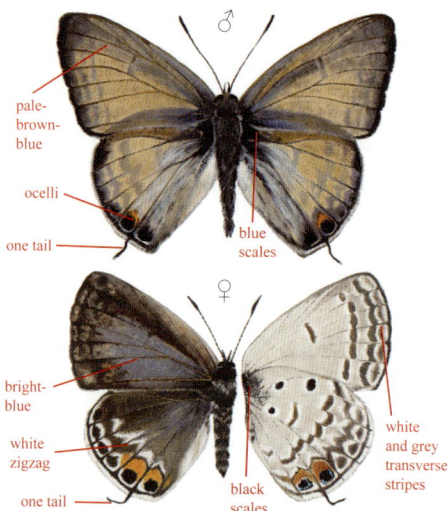

pale-brown-blue

ocelli

one tail

blue scales

bright-blue

white zigzag

one tail

black scales

white and grey transverse stripes

Distribution

This butterfly species occurs throughout the Afrotropical region including Comoros, Madagascar and Seychelles, where it occurs on Aldabra.

Biology and conservation

The Osiris Smokey Blue is not as widely collected in Seychelles as the other lycaenid species. Flight is brisk, generally close to the ground with the butterfly often settling. Both sexes mud-puddle. On Madagascar this species is a grassland butterfly (Lees et al. 2003).

Elsewhere, larvae feed on the Labiatae and Papilionaceae. This species occurs throughout the African continent and is not considered threatened.

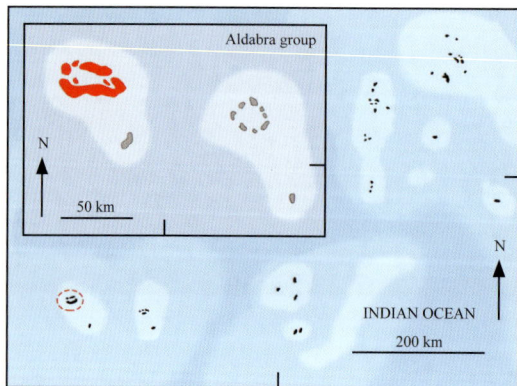

Aldabra group

N

50 km

N

INDIAN OCEAN

200 km

Above: *E. osiris*

Left: Seychelles distribution of *E. osiris*

83

South Africa

S. Woodhall

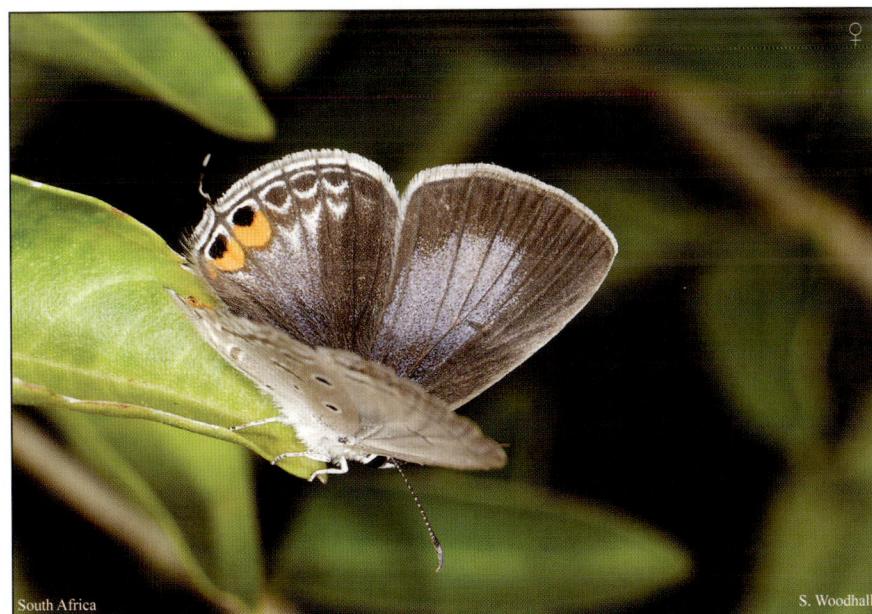

South Africa

S. Woodhall

Top: Male *E. osiris* perching

Bottom: Female *E. osiris* perching

References

Ackery PR, Vane-Wright RI (1984) *Milkweed Butterflies: Their Cladistics and Biology.* British Museum, Publication N° 893, Hong Kong.

Ackery PR, Smith CR, Vane-Wright RI (1995) *Carcasson's African Butterflies.* CSIRO, Australia.

Beamish T (1970) *Aldabra Alone.* George Allen & Unwin, London.

Berger LA (1962) Hesperiidae récoltés aux Seychelles. *Lambillionea* **LXII**:19–20.

Bernardi G (1954) Révision des Pierinae de la Malgache (Lépid. Pieridae). *Mémoires de l'Institut scientifique de Madagascar* **1954**:239–375.

Bernardi G (1996) Biogéographie et speciation des Lépidoptères Papilionidae, Pieridae, Danaidae et Acraeidae de Madagascar et des îles Voisines. *Biogéographie de Madagascar* **1996**:491–506.

Betts M (2000) Research officer's annual report, June 1999–July 2000 Aldabra Atoll. Unpublished report to Seychelles Island Foundation.

Betts M (2009) *Hypolimnas bolina* on Alphonse Island. *Phelsuma* **17**:57–58.

Bourquin O, Lawrence J, Hitchins P (2000) A note on the butterflies of Cousine Island, Seychelles. *Phelsuma* **8**:71–73.

Bowler J, Hunter J (1999) Aride Island Nature Reserve, Seychelles. Annual Report 1998. Unpublished report to Royal Society for Nature Conservation.

Braby MF, Vila R, Pierce NE (2006) Molecular phylogeny and systematics of the Pieridae (Lepidoptera: Papilionoidea): higher classification and biogeography. *Zoological Journal of the Linnean Society* **147**:239–275.

Brakefield PM, Larsen TB (1984) The evolutionary significance of dry and wet season forms in some tropical butterflies. *Biological Journal of the Linnean Society* **22**:1–12.

Braithwaite CJR (1984) Geology of Seychelles. In: *Biogeography and Ecology of the Seychelles Islands.* Stoddart DR (ed.) pp. 17–38. Junk Publishers, The Hague.

Carcasson RH (1964) New African butterflies. *Journal of the East African Natural History Society* **4(108)**:67–73.

Camoin GF, Montaggioni LF, Braithwaite CJR (2004) Late glacial to post glacial sea levels in the Western Indian Ocean. *Marine Geology* **206**:119–146.

Chiba H (2009) A revision of the subfamily Coeliadinae (Lepidoptera: Hesperiidae). *Bulletin of the Kitakyushu Museum of Natural History and Human History, Series A: Natural History* **7**:1–102.

Collins NM, Morris MG (1985) *Threatened Swallowtail Butterflies of the World: The IUCN Red Data Book*. IUCN, Gland, Switzerland.

Dall'Asta U, Turlin B (2009) Lépidoptères (Papillons de jour et Papillons de nuit). In: *La Fauna Terrestre de l'Archipel des Comoros*. Louette M, Meirte D, Jocqué R (ed.) pp. 276–304. Studies in Afrotropical Zoology, Nᵒ 293. Tervuren: MRAC.

Davies MA (2009) *Invasion Biology*. Oxford University Press, UK.

De Joannis J (1893) Trois macrolépidoptèrologique nouveaux des îles Seychelles. *Bulletin de la Société entomologique de France* **(1893)**:50–53.

De Joannis J (1894) Mission scientifique de Ch. Alluaud aux îles Seychelles. *Annales de la Société entomologique de France* **63**:425–438.

Evans WH (1937) *A Catalogue of the African Hesperiidae in the British Museum*. British Museum, London.

Fletcher T (1910) Lepidoptera, exclusive of the Tortricidae and Tineidae, with some remarks on their distribution and means of dispersal amongst the islands of the Indian Ocean. *Transactions of the Linnean Society, London* **13**:265–324.

Freitas AVL, Brown KS (2004) Phylogeny of the Nymphalidae (Lepidoptera). *Systematic Biology* **53**:363–383.

Friedmann F (2011) *Flore des Seychelles*. Publications Scientifiques de Muséum, Marseille.

Fryer JCF (1912) The Lepidoptera of the Seychelles and Aldabra, exclusive of the Orneodidae and Pterophoridae and of the Tortricina and Tineina. *Transactions of the Linnean Society, London* **15**:1–28.

Gerlach J (1997) *Seychelles Red Data Book 1997*. The Nature Protection Trust of Seychelles, Mahé.

Gerlach J (2008) Preliminary conservation status and needs of an oceanic island fauna: The case of Seychelles insects. *Journal of Insect Conservation* **12**:293–305.

Gerlach J (2012) Red listing reveals the true state of biodiversity: A comprehensive assessment of Seychelles biodiversity. *Phelsuma* **20**:9–22.

Gerlach J, Matyot P (2006) *Lepidoptera of the Seychelles Islands*. Backhuys Publishers, Leiden, The Netherlands.

Geyer C in Hübner J (1837) *Zuträge zur Sammlung exotischer Schmetterlinge 5*. Augsburg.

Gordon IJ, Edmunds M, Edgar JA, Lawrence J, Smith DA (2010) Linkage disequilibrium and natural selection for mimicry in the Batesian mimic *Hypolimnas misippus* (L.) (Lepidoptera: Nymphalidae) in the Afrotropics. *Biological Journal of the Linnean Society* **100**:180–194.

Hancock DL (1983) Classification of the Papilionidae (Lepidoptera): a phylogenetic approach. *Smithersia* **2**:1–48.

Henning GA (1986) A new species of *Acraea* F. (Lepidoptera: Nymphalidae) from South West Africa (Namibia) with revisional notes on the *Acraea horta* (L.) species group. *Journal of the Entomological Society of Southern Africa* **49**:29–37.

Henning GA, Henning SF, Joannou JG, Woodhall SE (1997) *Living Butterflies of Southern Africa: Biology, Ecology and Conservation Vol. 1.* Umdaus Press, South Africa.

Henri K, Milne GR, Shah NJ (2004) Costs of ecosystem restoration on islands in Seychelles. *Ocean and Coastal Management* **27**:409–428.

Hill MJ, Matyot P, Vel TM, Parr SJ, Shah NJ (2002) Marianne. *Atoll Research Bulletin* **495**:157–176.

Hill M, Currie D (2007) *Wildlife of Seychelles.* Collins Traveller's Guide, UK.

Holland WJ (1896) Lepidoptera from Aldabra, Seychelles, and other East African islands, collected by Dr. W.L. Abbott. *Proceedings of the United States National Museum* **1064**: 265–273.

Ladle RJ, Whittaker RJ (ed.) (2011) *Conservation Biogeography.* Wiley-Blackwell, UK.

Larsen TB (1996) *The Butterflies of Kenya and their Natural History.* Oxford University Press, UK.

Larsen TB (2005) *Butterflies of West Africa.* Apollo Books, Denmark.

Lawrence JM (2000a) Preliminary evaluation of the composition, distribution and faunal affinities of the butterflies of the Seychelles archipelago. *Metamorphosis* **11**:174–184.

Lawrence JM (2000b) Meso-distribution, abundance and phenotypic variation in *Hypolimnas misippus* (Lepidoptera: Nymphalidae) on Cousine Island, Seychelles. *Metamorphosis* **11**:146–153.

Lawrence JM (2004a) Ecology and biology of a Seychelles island *Borbo gemella* (Lepidoptera: Hesperiinae) metapopulation. *Phelsuma* **12**:153–157.

Lawrence JM (2004b) Lepidoptera of Bird Island, Seychelles. *Metamorphosis* **15**:77–79.

Lawrence JM (2005) Lepidoptera of Cousine Island, Seychelles. *Phelsuma* **13**:94–101.

Lawrence JM (2009a) Comparative biometrics of a Seychelles island *Borbo gemella* (Lepidoptera: Hesperiinae) metapopulation. *Phelsuma* **17**:46–49.

Lawrence JM (2009b) *Euploea rogeri* (Lepidoptera: Danainae), a little known Seychelles butterfly. *Phelsuma* **17**:53–56.

Lawrence JM (2009c) Habitat restoration and the decline of the African Grass Blue, *Zizeeria knysna* (Lepidoptera: Lycaenidae), on a small Seychelles island. *Metamorphosis* **20**:131–140.

Lawrence JM (2010) A short note on the Striped Policeman butterfly, *Coeliades forestan* (Lepidoptera: Hesperiidae) in Seychelles. *Phelsuma* **18**:90–94.

Lawrence JM (2011) A further note on *Hypolimnas bolina* in Seychelles. *Phelsuma* **19**: 41–42.

Lees DC, Kremen C, Raharitsimba T (2003) Classification, diversity, and endemism of the butterflies (Papilionoidea and Hesperioidea): A revised species checklist. In: *The Natural History of Madagascar*. Goodman SM, Benstead JP (ed.) pp. 762–793. The University of Chicago Press, Chicago.

Legrand H (1959) Note sur la sous-espèce *nana* Ch. Oberthür de *Papilio phorbanta* Linné des îles Seychelles (Lep. Papilionidae). *Bulletin de la Société Entomologique de France* **64**:121–123.

Legrand H (1965) Lépidoptères des îles Seychelles et d'Aldabra. *Mémoires du Muséum National D'Historie Naturelle* **37**:1–210.

Lionnet G (1970) Note on the Lepidoptera of Astove Atoll. *Atoll Research Bulletin* **136**:113–114.

Matyot P (2002) Observations on some rail/insect interactions on Aldabra. *Birdwatch* **43**.

Martiré D, Rochat J (2008) *Les Papillons de La Réunion*. Biotope, France.

Nazari V, Larsen TB, Lees DC, Brattström O, Bouyer T, Van De Poel G, Hebert PDN (2011) Phylogenetic systematics of *Colotis* and associated genera (Lepidoptera: Pieridae): evolutionary and taxonomic implications. *Journal of Zoological Systematics and Evolutionary Research* **49**:204–215.

Nelson SM (2007) Butterflies (Papilionoidea and Hesperioidea) as potential ecological indicators of riparian quality in the semi-arid western United States. *Ecological Indicators* **7**:469–480.

New TR (2014) *Lepidoptera and Conservation*. Wiley-Blackwell, UK.

New TR, Collins NM (1991) *Swallowtail Butterflies: An Action Plan for their Conservation*. IUCN, Gland, Switzerland.

O'Dowd DJ, Green PT, Lakes PS (2003) Invasional 'meltdown' on an oceanic island. *Ecological Letters* **6**:445–461.

Ogilvie-Grant WR (1903) Lepidoptera I. – Rhopalocera. In: *The Natural History of Sokotra and Abd al Kuri*. Forbes HO (ed.) pp. 293–318. Henry Young & Sons, London.

Paulian R (1956) Insectes: Lépidoptères; Danaidae, Nymphalidae, Acraeidae. *Faune de Madagascar* **2**:1–102.

Paulian R, Viette P (1968) Insectes: Lépidoptères; Papilionidae. *Faune de Madagascar* **27**:1–97.

Pierre J, Bernaud D (2013) *Butterflies of the World, Part 39:* Acraea *subgenus* Acraea. Goecke & Evers, Keltern.

Piggott CJ (1968) *A Soil Survey of Seychelles*. Technical Bulletin No. 2. Land Resource Division, Directorate of Overseas Surveys, Tolworth, Surrey, UK.

Procter J (1984a) Vegetation of the granitic islands of the Seychelles. In: *Biogeography and Ecology of the Seychelles Islands*. Stoddart DR (ed.) pp. 193–207. Junk Publishers, The Hague.

Procter J (1984b) Floristics of the granitic islands of the Seychelles. In: *Biogeography and Ecology of the Seychelles Islands*. Stoddart DR (ed.) pp. 209–220. Junk Publishers, The Hague.

Primack RB (2006) *Essentials of Conservation Biology*. Sinauer, Sunderland.

Robertson SA (1989) *Flowering Plants of Seychelles*. Royal Botanic Gardens, Kew, UK.

Samways MJ (2005) *Insect Diversity Conservation*. Cambridge University Press, UK.

Samways MJ, McGeoch MA, New TR (2010a) *Insect Conservation: A Handbook of Approaches and Methods*. Oxford University Press, New York.

Samways MJ, Hitchins P, Bourquin O, Henwood J (2010b) *Tropical Island Recovery, Cousine Island, Seychelles*. Wiley-Blackwell, Oxford, UK.

Shah NJ (2001) Eradication of alien predators in Seychelles: an example of conservation action on tropical islands. *Biodiversity and Conservation* **10**:1219–1220.

Shah NJ (2006) Ecological restoration of islands in the Seychelles. *Conservation Evidence* **3**:1–2.

Shields O (1989) World numbers of butterflies. *Journal of the Lepidopterists' Society* **43**: 178–183.

Smith DAS, Lushai G, Allen JA (2005) A classification of *Danaus* butterflies (Lepidoptera: Nymphalidae) based upon data from morphology and DNA. *Zoological Journal of the Linnean Society* **144**:191–212.

Stoddart DR (1984) Impact of man in the Seychelles. In: *Biogeography and Ecology of the Seychelles Islands*. Stoddart DR (ed.) pp. 642–654. Junk Publishers, The Hague.

Tingay P (2010) *Globetrotter Travel Guide: Seychelles*. New Holland, London.

Turlin B (1995) Faune Lépidoptèrologique de l'archipel des Comoros (6). *Lambillionea* **XCV**:197–210.

Viette P (1957) Lépidoptères (excepté les Tordeuses et les Géométrides). *Mémoires de l'Institut scientifique de Madagascar* **8**:137–226.

Viette P (1958) Insectes: Lépidoptères; Hesperiidae. *Faune de Madagascar* **3**:1–85.

Vinson J (1938) Catalogue of the Lepidoptera of the Mascarene Islands. *Mauritius Institute Bulletin* **1**:1–69.

Wahlberg N, Brower AVZ, Nylin S (2005) Phylogenetic relationships and historical biogeography of tribes and genera in the subfamily Nymphalinae (Lepidoptera: Nymphalidae). *Biological Journal of the Linnean Society* **86**:227–251.

Wahlberg N, Leneveu J, Kodandaramaiah U, Pena C, Nylin S, Freitas AVL, Brower AVZ (2009) Nymphalid butterflies diversify following near demise at the Cretaceous/Tertiary boundary. *Proceedings of the Royal Society B: Biological Sciences* **276**:4295–4302.

Walsh RPD (1984) Climate of the Seychelles. In: *Biogeography and Ecology of the Seychelles Islands*. Stoddart DR (ed.) pp. 39–62. Junk Publishers, The Hague.

Warren AD, Ogawa JR, Brower AVZ (2008) Phylogenetic relationships of subfamilies and circumscription of tribes in the family Hesperiidae (Lepidoptera: Hesperioidea). *Cladistics* **24**:642–676.

Warren AD, Ogawa JR, Brower AVZ (2009) Revised classification of the family Hesperiidae (Lepidoptera: Hesperioidea) based on combined molecular and morphological data. *Systematic Entomology* **34**:467–523.

Whittaker RJ, Fernández-Palacios JM (2007) *Island Biogeography: Ecology, Evolution, and Conservation*. Oxford University Press, UK.

Williams JR (2007) *Butterflies of Mauritius*. Bioculture Press, Mauritius.

Yata O (1994) A revision of the Old World species of the genus *Eurema* Hübner (Lepidoptera, Pieridae): Part IV. Description of the *hecabe* group (part). *Bulletin of the Kitakyushu Museum of Natural History* **13**:59–105.

Zamin TJ, Baillie JM, Miller RM, Rodriguez JP, Ardid A, Collen B (2009) National Red Listing beyond 2010 target. *Conservation Biology* **24**:1012–1020.

D. Lawrence

Above: Assumption

Glossary

Abiotic: Non-living.

Anthropogenic: Produced or caused by man.

Basal: The area at or adjacent to the base of a structure. For example, the area on a wing that is closest to the thorax.

Batesian mimicry: The mimicking of a distasteful or poisonous species by an edible species.

Biotic: Living.

Biotope: A region or area (e.g. woodland, tropical forest, cliff) that is distinguished by particular biotic and abiotic environmental conditions.

Character: A feature or a trait that differentiates one member of a population or taxon from a member of another population or taxon.

Chitin: A long chain of polymer N-acetylglucamine units, the chief polysaccharide in the exoskeleton of arthropods and in fungal cell walls.

Clade: A group containing the set of all organisms descended from a common ancestor.

Compound eye: An eye that is made up of several elements. It consists of hundreds of tiny separate elements, called ommatidia, each with its own visual cells. The primary eyes in an invertebrate.

Coxa: The first segment of the leg.

Cremaster: The constricted anal segment of the pupa that is attached to the silk pad spun by the larva before pupation.

Cremaster hooks: Tiny hooks that attach the pupa to the silk pad.

Crepuscular: Active during twilight.

Crochets: Tiny sclerotised hooks at the end of the prolegs in the larva.

Cuticle: The non-cellular skin of an invertebrate that is made of chitin and protein.

Cryptic: Protective colouring making concealment easier.

Description: In taxonomy, a formal statement of the characters that define a taxon.

Dimorphism: The occurrence of a species in two or more forms. Where these forms are seasonal its known as seasonal dimorphism, and where males and females differ it is known as sexual dimorphism.

Discal cell: The large central area of a wing surrounded by veins.

Dorsal: Upper-surface.

Endemic/Endemism: Naturally occurring in a specified geographical area, e.g. a country.

Exoskeleton: The hardened external integument in an arthropod.

Extirpation: Localised extinction of a taxon.

Eyespot: A coloured spot, ringed with contrasting colours, found on the wing. Also called an ocellus.

Family: The taxonomic rank above genus and below order. Usually comprises several subfamilies.

Femur: The third segment of an insect's leg.

Food plant: The host plant on which larvae feed.

Form: Individuals of a species that differ from other individuals of the same species in a consistent way. A form may be under genetic or environmental control. Form names are not recognised under the International Code of Zoological Nomenclature, and are only used in this book where they are well known (e.g. for *Hypolimnas misippus*). Intermediates between different forms are common.

Genus: The taxonomic rank above species. A group of closely related species. Plural is genera.

Girdle: A silk attachment strand found on the pupa of the Hesperiidae, Pieridae, Papilionidae and some Lycaenidae. It is not found in the Nymphalidae.

Ground colour: The prominent colour of a wing.

Hair-pencil: A dense tuft of long hairs on the wings or body of a male butterfly that is used to dispense pheromones during courtship.

Haustellum: The coiled proboscis that is used for feeding.

Humeral vein: A short vein near the base of the hindwing.

Hyaline: Transparent. Due to the absence of scales. Usually associated with clear patches on the wing.

Intraspecific: Within a species.

Irrorated: Very finely speckled with small dots, streaks or lines.

Instar: The larval stages between successive moults.

Labial palpi: Paired appendages forming part of the insect's mouthparts.

Marginal: Describing the features found on the outer margins of the wings.

Metamorphosis: The transformation from larval to adult form.

Micropyle: The small opening at the top of an egg through which sperm enters during fertilisation. Also used as the respiratory opening during egg development.

Migration: The mass movement of animals in one direction at one time.

Mimicry: The resemblance of one species to another species.

Morphology: The description of the structural characteristics of a taxon or individual.

Model: A distasteful species that a palatable species resembles.

Moult: The growth replacement of the cuticle in the immature stages of an insect. Also known as ecdysis.

Mud-puddling: A behaviour were a butterfly sips salts and minerals from moist mud.

Müllerian mimicry: Where an unpalatable species mimics another unpalatable species, with both species benefitting from the resemblance.

Myrmecophilous: Associated with ants.

Nominate: The first of two or more subspecies named.

Ocellus: 1) The small simple eye of an adult insect; 2) Alternative term for wing eyespot. The plural is ocelli.

Oviposition: To lay to an egg.

Pheromone: A chemical secreted by an individual to attract another individual. Most commonly associated with a male attracting a female. For myrmecophilous species pheromones are used to attract ants.

Polymorphism: The occurrence of more than one morphological form in a particular species.

Polyphagous: Feeding on more than one plant species.

Prolegs: Fleshy leg-like organs on abdominal segments three to six and 10 in the larvae.

Sex-brand: A patch of dense anacondria scales (i.e. specialised wing scales that produce pheromones) on the wings of some male butterflies.

Simple eye: An invertebrate eye that only has a single lens, usually found in the larvae.

Species: A population of interbreeding organisms that share a common mate-recognition system. It is the primary biological unit.

Spiracle: Paired openings on the sides of an insect's abdomen, used for breathing.

Subspecies: One or more taxonomically distinct populations within a species. Each population is geographically isolated from the other.

Tarsus: The last segment of an insect's leg. Plural is tarsi.

Taxon: A taxonomic group that is sufficiently distinct to be worthy of being distinguished by a name and to be ranked in a definite category. Plural is taxa.

Taxonomy: The science of classifying organisms.

Tibia: The fourth segment of an insect's leg.

Trochanter: The second segment of an insect's leg, placed between the coxa and femur.

Venation: The arrangement of the veins on an insect's wing.

Ventral: Under-surface.

Wingspan: The size of an insect measured across the forewings from apex to apex of a set specimen.

Top left: *Junonia orithya madagascariensis*; **Top right**: *Borbo gemella*
Second row left: *Zizula hylax*; **Second row right**: *Hypolimnas misippus*
Third row left: *Danaus chrysippus orientis*; **Third row right**: *Danaus dorippus dorippus*
Bottom left: *Zizeeria knysna*; **Bottom right**: *Acraea neobule*

Plate 1: Hesperiidae

Plate 2: Papilionidae

Papilio phorbanta nana ♂
page 38

Papilio phorbanta nana ♀
page 38

Papilio dardanus ♂
page 39

Plate 3: Pieridae

Plate 4: Pieridae

Belenois grandidieri ♀
page 45

Belenois grandidieri ♂
page 45

Belenois aldabrensis ♀
page 46

Belenois aldabrensis ♂
page 46

Colotis evanthe evanthides ♂
page 47

Colotis evanthe evanthides ♀
page 47

Colotis evanthe evanthides ♂
page 47

Plate 5: Nymphalidae

Plate 6: Nymphalidae

Euploea mitra ♂
pages 53–54

Euploea mitra ♀
pages 53–54

Euploea rogeri dorsal surface ♀
page 52

Euploea rogeri ventral surface ♀
page 52

Plate 7: Nymphalidae

Plate 8: Nymphalidae

Melanitis leda helena ♀
pages 59–60

Melanitis leda helena ♂
pages 59–60

Plate 9: Nymphalidae

Melanitis leda helena ♂
pages 59–60

Melanitis leda helena ♂
pages 59–60

Plate 10: Nymphalidae

Hypolimnas misippus ♂
pages 61–62

Hypolimnas misippus ♀
pages 61–62

Hypolimnas misippus ♀
pages 61–62

Hypolimnas misippus ♀
pages 61–62

Plate 11: Nymphalidae

Hypolimnas bolina jacintha ♂
pages 63–64

Hypolimnas bolina jacintha ♀
pages 63–64

Plate 12: Nymphalidae

Vanessa cardui ♂
page 72

Junonia rhadama ♀
pages 65–66

Junonia rhadama ♂
pages 65–66

Junonia rhadama ventral surface
pages 65–66

Plate 13: Nymphalidae

Junonia hierta cebrene ♀
page 71

Junonia hierta cebrene ♂
page 71

*Junonia orithya
madagascariensis* ♀
pages 69–70

*Junonia orithya
madagascariensis* ♂
pages 69–70

Junonia oenone epiclelia ♀
pages 67–68

Junonia oenone epiclelia ♂
pages 67–68

Junonia oenone epiclelia ♀
pages 67–68

Plate 14: Lycaenidae

Index to Scientific Names

A

Acraea
   neobule 55, 94, 107
   ranavalona 56, 107
   terpsicore 55
      legrandi 55
Amauris
   niavius
      dominicanus 51, 103

B

Belenois
   aldabrensis 22, 23,45, 46, 101
   grandidieri 45, 101
Borbo
   borbonica 22, 23, 95
      borbonica 34, 95
      morella 22, 23, 27, 33, 34, 95
   gemella 35, 94

C

Catopsilia
   florella 41, 42, 99
Coeliades
   forestan
      arbogastes 25, 29, 30, 95
      forestan 25, 30, 95
Colotis
   etrida 40
   evanthe 19
      evanthides 23, 40, 47, 101

D

Danaus
   chrysippus 20, 21, 50, 62
      orientis 49, 94, 103
   dorippus 20, 21, 62
      dorippus 50, 94, 103

E

Eagris
   sabadius 20
      aldabranus 20, 23, 31, 32, 95
      maheta 20, 23, 32, 95

Euchrysops
   malathana 73
   osiris 73, 83, 84, 121
Euploea
   euphon 22, 52
   mitra 17, 19, 22, 23, 52, 53, 54, 105
   rogeri 52, 105
Eurema 19, 40
   brigitta 20
      pulchella 20, 43, 99
   desjardinsii 40
   floricola 20
      aldabrensis 20, 23, 44, 99

H

Hypolycaena
   philippus
      ramonza 73, 74, 121
Hypolimnas
   bolina
      jacintha 63, 64, 115
   misippus 19, 20, 21, 61, 62, 63, 94, 113
      f. inaria 20, 21, 61
      f. misippus 20, 21, 61

J

Junonia
   hierta
      cebrene 70, 71, 119
   oenone
      epiclelia 67, 68, 119
   orithya
      madagascariensis 69, 70, 94, 119
   rhadama 19, 65, 66, 117

L

Lampides
   boeticus 73, 75, 76, 121
Leptotes
   pirithous 73, 77, 78, 121

M

Melanitis
   leda 19
      helena 59, 60, 109, 111

D. Lawrence

Above: Farquhar Atoll

Index to Common Names